Nonlinear Functional Analysis and Its Applications

Nonlinear Functional Analysis and Its Applications

Editor

Radu Precup

MDPI • Basel • Beijing • Wuhan • Barcelona • Belgrade • Manchester • Tokyo • Cluj • Tianjin

Editor
Radu Precup
Babeş-Bolyai University
Romania

Editorial Office
MDPI
St. Alban-Anlage 66
4052 Basel, Switzerland

This is a reprint of articles from the Special Issue published online in the open access journal *Mathematics* (ISSN 2227-7390) (available at: https://www.mdpi.com/journal/mathematics/special_issues/Nonlinear_Function).

For citation purposes, cite each article independently as indicated on the article page online and as indicated below:

LastName, A.A.; LastName, B.B.; LastName, C.C. Article Title. *Journal Name* **Year**, *Volume Number*, Page Range.

ISBN 978-3-0365-0240-3 (Hbk)
ISBN 978-3-0365-0241-0 (PDF)

Contents

About the Editor . . . . **vii**

Preface to "Nonlinear Functional Analysis and Its Applications" . . . . **ix**

Jean Mawhin
Variations on the Brouwer Fixed Point Theorem:
A Survey
Reprinted from: *Mathematics* **2020**, *8*, 501, doi:10.3390/math8040501 . . . . **1**

Dumitru Motreanu, Angela Sciammetta and Elisabetta Tornatore
A Sub-Supersolution Approach for Robin Boundary Value Problemswith Full Gradient Dependence
Reprinted from: *Mathematics* **2020**, *8*, 658, doi:10.3390/math8050658 . . . . **15**

Donal O'Regan
The Topological Transversality Theorem for Multivalued Maps with Continuous Selections
Reprinted from: *Mathematics* **2019**, *7*, 1113, doi:10.3390/math7111113 . . . . **29**

Biagio Ricceri
A Class of Equations with Three Solutions
Reprinted from: *Mathematics* **2020**, *8*, 478, doi:10.3390/math8040478 . . . . **35**

Biagio Ricceri
Correction: Ricceri, B. A Class of Equations with Three Solutions. *Mathematics* 2020, *8*, 478
Reprinted from: *Mathematics* **2021**, *9*, 101, doi:10.3390/math9010101 . . . . **43**

Rodrigo López Pouso, Radu Precup and Jorge Rodríguez-López
Positive Solutions for Discontinuous Systems via a Multivalued Vector Version of Krasnosel'skiĭ's Fixed Point Theorem in Cones
Reprinted from: *Mathematics* **2019**, *7*, 451, doi:10.3390/math7050451 . . . . **45**

Xiaoyan Shi, Yulin Zhao and Haibo Chen
Existence of Solutions for Nonhomogeneous Choquard Equations Involving p-Laplacian
Reprinted from: *Mathematics* **2019**, *7*, 871, doi:10.3390/math7090871 . . . . **61**

Binghua Jiang, Huaping Huang and Wei-Shih Du
New Generalized Mizoguchi-Takahashi's Fixed Point Theorems forEssential Distances and e^0-Metrics
Reprinted from: *Mathematics* **2019**, *7*, 1224, doi:10.3390/math7121224 . . . . **79**

Jiunn-Shiou Fang, Jason Sheng-Hong Tsai, Jun-Juh Yan, Chang-He Tzou and Shu-Mei Guo
Design of Robust Trackers and Unknown Nonlinear Perturbation Estimators for a Class of Nonlinear Systems: HTRDNA Algorithm for Tracker Optimization
Reprinted from: *Mathematics* **2019**, *7*, 1141, doi:10.3390/math7121141 . . . . **95**

Anibal Coronel, Francisco Novoa-Muñoz, Ian Hess and Fernando Huancas
Analysis of a SEIR-KS Mathematical Model For Computer Virus Propagation in a Periodic Environment
Reprinted from: *Mathematics* **2020**, *8*, 761, doi:10.3390/math8050761 . . . . **115**

About the Editor

Radu Precup, Professor, received his Ph.D. degree in Mathematics from Babeş-Bolyai Unversity of Cluj-Napoca, Romania, in 1985 and held a postdoctoral BGF fellowship at Paris 6 University between October 1990 and June 1991. He is currently Full Professor in the Department of Mathematics at Babeş-Bolyai University. Dr. Precup's research interests include nonlinear functional analysis, nonlinear ordinary and partial differential equations, and biomathematics. He authored over 150 research papers and the books Methods in Nonlinear Integral Equations (2002, Springer), Theorems of Leray-Schauder Type and Applications (with D. O'Regan, 2001, CRC), Linear and Semilinear Partial Differential Equations (2013, De Gruyter), and Ordinary Differential Equations (2018, De Gruyter).

Preface to "Nonlinear Functional Analysis and Its Applications"

Originally, functional analysis was that branch of mathematics capable of investigating in an abstract way a series of linear mathematical models from science. The study of these linear models—in fact, only first approximations of real models—proved insufficient, so the theory had to be extended to be able to describe the nonlinear phenomena themselves. In this way nonlinear functional analysis was born and continues to develop, becoming a vast and fascinating field of mathematics, with deep applications to increasingly complex problems in physics, biology, chemistry, and economics.

This book consists of nine papers covering a number of basic ideas, concepts, and methods of nonlinear analysis, as well as some current research problems. Thus, the reader is introduced to the fascinating theory around Brouwer's fixed point theorem, which is the basis of important extensions to infinitely dimensional spaces with numerous applications to boundary value problems for various classes of ordinary and partial differential equations. New results for nonstandard elliptic equations obtained with methods such as the technique of upper and lower solutions, advanced methods of critical point theory, and minimax techniques are then presented. The reader is also introduced to Granas' theory of topological transversality, an alternative to the theory of topological degree. Several contributions address current research issues, such as the problem of discontinuous term equations, results of metric fixed point theory, robust tracker design problems for various classes of nonlinear systems, or the problem of periodic solutions in computer virus propagation models.

I would like to particularly thank Professor Jean Mawhin, Professor Dumitru Motreanu, Professor Donal O'Regan, and Professor Biagio Ricceri, who have positively answered our invitation to contribute a paper to this Special Issue. I am sure that their extremely valuable papers will interest the readers and will stimulate new research work in this direction. I would also like to thank the other contributors for their articles that open new perspectives over some specific problems and applications.

Finally, I would like to thank the editors of the journal *Mathematics*, particularly Assistant Editor Grace Du and Marketing Assistant Rainy Han, for their great support throughout the editing process of the Special Issue for Mathematics and its present MDPI Reprint Book.

Radu Precup
Editor

mathematics

MDPI

Review

Variations on the Brouwer Fixed Point Theorem: A Survey

Jean Mawhin

Institut de Recherche en Mathématique et Physique, Université Catholique de Louvain, chemin du Cyclotron, 2, 1348 Louvain-la-Neuve, Belgium; jean.mawhin@uclouvain.be

Received: 25 February 2020; Accepted: 21 March 2020; Published: 2 April 2020

Abstract: This paper surveys some recent simple proofs of various fixed point and existence theorems for continuous mappings in $\mathbb{R}^n$. The main tools are basic facts of the exterior calculus and the use of retractions. The special case of holomorphic functions is considered, based only on the Cauchy integral theorem.

Keywords: Brouwer fixed point theorem; Hamadard theorem; Poincaré–miranda theorem

MSC: 55M20; 54C15; 30C15

---

1. Introduction

The Bolzano theorem for continuous functions $f : [a, b] \subset \mathbb{R} \to \mathbb{R}$, which states that *$f$ has a zero in $[a, b]$ if $f(a)f(b) \leq 0$*, was first proved in 1817 by Bolzano [1] and, independently and differently in 1821 by Cauchy [2]. Its various proofs are not very long, and depend only upon the order and completeness properties of $\mathbb{R}$. A consequence of the Bolzano theorem applied to $I - T$ is that *$T : [-R, R] \to \mathbb{R}$, continuous, has a fixed point in $[-R, R]$ if $T(-R) \in [-R, R]$ and $T(R) \in [-R, R]$*. This is the case if $T : [-R, R] \to [-R, R]$.

As $[-R, R]$ is the closed ball of center 0 and radius R in $\mathbb{R}$, a natural question is to know if, B_R denoting the closed ball $B_R \subset \mathbb{R}^n$ of center 0 and radius $R > 0$, *any continuous mapping $T : B_R \to \mathbb{R}^n$ such that $T(\partial B_R) \subset B_R$ has a fixed point*, and, in particular, if *any continuous mapping $T : B_R \to B_R$ has a fixed point*. The answer is yes, and the first result, usually called the Rothe fixed point theorem (FPT), is more correctly referred as the Birkhoff–Kellog FPT, and the second one as the Brouwer FPT.

Many different proofs of those results have been given since the first published one of the Brouwer FPT by Hadamard in 1910 [3]. Brouwer's original proof [4], published in 1912, was topological and based on some fixed point theorems on spheres proved with the help of the topological degree introduced in the same paper. The Birkhoff–Kellogg FPT was first proved by Birkhoff and Kellogg in 1922 [5]. Its standard name Rothe FPT refers to its extension to Banach spaces by Rothe [6] in 1937.

The existing proofs use ideas from various areas of mathematics such as algebraic topology, combinatorics, differential topology, analysis, algebraic geometry, and even mathematical economics. A survey and a bibliography can be found in [7]. Even for $n = 2$, they cease to be elementary and/or can be technically complicated. The aim of this paper is to survey recent results on some elementary approaches to the Birkhoff–Kellogg and Brouwer FPT, and on how to deduce from them in a simple and systematic way other fixed point and existence theorems for mappings in $\mathbb{R}^n$. Recall that these results, combined with basic facts of functional analysis, are fundamental in obtaining useful extensions to some classes of mappings in infinite-dimensional normed spaces.

After recalling the simple concept of curvilinear integral in $\mathbb{R}^2$, we first propose in Section 2 an elementary proof of the Birkhoff–Kellogg FPT for $n = 2$, based upon such integrals. As the extension to arbitrary n, using differential $(n-1)$ forms in $\mathbb{R}^n$, leads to very cumbersome computations, we adopt in Section 3 a variant given in [8], using differential n-forms, which in dimension n happens to be significantly simpler than the direct extension of the approach of Section 2.

The generalizations of the Birkhoff–Kellogg and Brouwer FPT to a closed ball in $\mathbb{R}^n$ and their homeomorphic images are stated in Section 4. After the concepts of retract and retraction are introduced, the Leray–Schauder–Schaefer FPT on a closed ball is deduced from the Brouwer FPT, whose statement is also extended to retracts of a closed ball in $\mathbb{R}^n$. Finally, the equivalence of the Birkhoff–Kellogg and Brouwer FPT on a closed ball is established.

The Brouwer FPT and retractions are then used in Section 5 to prove, in a very simple and unified way inspired by the approach of [9], several conditions for the existence of zeros continuous mappings in $\mathbb{R}^n$, namely the Poincaré–Bohl theorem on a closed ball, the Hadamard theorem on a compact convex set, the Poincaré–Miranda theorem on a closed n-interval, and the Hartman–Stampacchia theorem on variational inequalities.

Finally, in Section 6, following the method introduced in [10], simple versions of the Cauchy integral theorem provide criterions for the existence of zeros of a holomorphic function in same spirit of the approach in Section 2. They allow very simple proofs of the Hadamard and Poincaré–Miranda theorems and of the Birkhoff–Kellogg and Brouwer FPT for holomorphic functions.

2. A Proof the Birkhoff–Kellogg Theorem on a Closed Disc Based on Curvilinear Integrals

Let $D \subset \mathbb{R}^2$ be open and nonempty and let $\langle \cdot, \cdot \rangle$ denote the usual inner product in $\mathbb{R}^2$. Given $f = (f_1, f_2) : D \to \mathbb{R}^2$, $x \mapsto f(x)$ and $\varphi = (\varphi_1, \varphi_2) : [a,b] \to D, t \mapsto \varphi(t)$ of class C^1, we consider the corresponding **curvilinear integral** defined by $\int_a^b \langle f(\varphi(t)), \varphi'(t) \rangle dt$ where $'$ denotes the derivative with respect to t.

The following result is fundamental for our proof of the Birkhoff–Kellogg FPT on a closed disc.

Lemma 1. *If $f = (f_1, f_2) : D \to \mathbb{R}^2$ is of class C^1 and such that $\partial_1 f_2 = \partial_2 f_1$ and if $\Phi : [a,b] \times [0,1] \to D$ is of class C^2 and such that $\Phi(b,\lambda) = \Phi(a,\lambda)$ for all $\lambda \in [0,1]$, then $\lambda \to \int_a^b \langle f(\Phi(t,\lambda)), \partial_t \Phi(t,\lambda) \rangle dt$ is constant on $[0,1]$.*

Proof. It suffices to prove that $\partial_\lambda \int_a^b \langle f(\Phi(t,\lambda)), \partial_t \Phi(t,\lambda) \rangle dt = 0$ for all $\lambda \in [0,1]$. We have, with differentiation under integral sign easily justified and the use of assumptions, the Schwarz theorem and the fundamental theorem of calculus, and omitting the arguments (t,λ) for the sake of brevity

$$
\begin{aligned}
&\partial_\lambda \int_a^b \langle f(\Phi), \partial_t \Phi \rangle dt = \int_a^b \partial_\lambda [\langle f(\Phi), \partial_t \Phi \rangle] dt \\
= &\int_a^b \{ \langle \partial_\lambda [f(\Phi)], \partial_t \Phi \rangle + \langle f(\Phi), \partial_\lambda \partial_t \Phi \rangle \} dt \\
= &\int_a^b \left[\left\langle \sum_{j=1}^2 \partial_j f(\Phi) \partial_\lambda \Phi_j, \partial_t \Phi \right\rangle + \langle f(\Phi), \partial_t \partial_\lambda \Phi \rangle \right] dt \\
= &\int_a^b \left[\sum_{k=1}^2 \sum_{j=1}^2 \partial_j f_k(\Phi) \partial_\lambda \Phi_j \partial_t \Phi_k + \langle f(\Phi), \partial_t \partial_\lambda \Phi \rangle \right] dt \\
= &\int_a^b \left[\sum_{k=1}^2 \sum_{j=1}^2 \partial_k f_j(\Phi) \partial_t \Phi_k \partial_\lambda \Phi_j + \langle f(\Phi), \partial_t \partial_\lambda \Phi \rangle \right] dt \\
= &\int_a^b \left[\sum_{j=1}^2 \partial_t f_j(\Phi) \partial_\lambda \Phi_j + \langle f(\Phi), \partial_t \partial_\lambda \Phi \rangle \right] dt \\
= &\int_a^b \{ \langle \partial_t [f(\Phi)], \partial_\lambda \Phi \rangle + \langle f(\Phi), \partial_t \partial_\lambda \Phi \rangle \} dt \\
= &\int_a^b \partial_t [\langle f(\Phi), \partial_\lambda \Phi \rangle] dt = f(\Phi(b,\lambda)) - f(\Phi(a,\lambda)) = 0.
\end{aligned}
$$

□

Let $B_R := \{x \in \mathbb{R}^2 : |x| \leq R\}$, with $|x|$ the Euclidian norm. We prove the **Birkhoff–Kellogg FPT on a closed disc.**

Theorem 1. *Any continuous mapping $T : B_R \to \mathbb{R}^2$ such that $T(\partial B_R) \subset B_R$ has a fixed point in B_R.*

Proof. Assume that T has no fixed point in B_R. Then, $|y - T(y)| > 0$ for all $y \in \partial B_R$, and, as $T(\partial B_R) \subset B_R$, $|y - \lambda T(y)| \geq |y| - \lambda|T(y)| \geq 1 - \lambda > 0$, for all $(y, \lambda) \in \partial B_R \times [0, 1)$. Similarly, $\lambda y - T(\lambda y) \neq 0$ for all $(y, \lambda) \in \partial B_R \times [0, 1]$. As T is continuous, there exists $\delta > 0$ such that $|y - \lambda T(y)| \geq \delta$ and $|\lambda y - T(\lambda y)| \geq \delta$ for all $(y, \lambda) \in \partial B_R \times [0, 1]$. From the Weierstrass approximation theorem, there is a polynomial $P : \mathbb{R}^2 \to \mathbb{R}^2$ such that $|T(y) - P(y)| \leq \frac{\delta}{2}$ for all $y \in B_R$. Consequently, letting $F(y, \lambda) := y - \lambda P(y)$ and $G(y, \lambda) := \lambda y - P(\lambda y)$, we have, for all $(y, \lambda) \in \partial B_R \times [0, 1]$, $|F(y, \lambda)| \geq \frac{\delta}{2}$ and $|G(y, \lambda)| \geq \frac{\delta}{2}$. Hence, there exists an open neighborhood Δ of ∂B_R such that $F(y, \lambda) \neq 0$ and $G(y, \lambda) \neq 0$ for all $(y, \lambda) \in \Delta \times [0, 1]$. If

$$f_1 : \mathbb{R}^2 \setminus \{0\} \to \mathbb{R},\ x \mapsto -|x|^{-2} x_2,\ \ f_2 : \mathbb{R}^2 \setminus \{0\} \to \mathbb{R}, x \mapsto |x|^{-2} x_1,$$

then $\partial_2 f_1(x) = |x|^{-4}(x_2^2 - x_1^2) = \partial_1 f_2(x)$. If $\gamma_R : [0, 2\pi] \to \mathbb{R}^2,\ t \mapsto R(\cos t, \sin t)$ is a parametric representation of ∂B_R, so that $\gamma_R(0) = \gamma_R(2\pi)$, it follows from Lemma 1 that the integrals

$$\int_0^{2\pi} \langle f[F(\gamma_R(t), \lambda)], \partial_t F(\gamma_R(t), \lambda)\rangle dt \text{ and } \int_0^{2\pi} \langle f[G(\gamma_R(t), \lambda)], \partial_t G(\gamma_R(t), \lambda)\rangle dt$$

are constant for $\lambda \in [0, 1]$. Hence, noticing that $F(\cdot, 1) = G(\cdot, 1) = I - P$,

$$\begin{aligned} &\int_0^{2\pi} \langle f[F(\gamma_R(t), 0)], \partial_t F(\gamma_R(t), 0)\rangle dt = \int_0^{2\pi} \langle f[F(\gamma_R(t), 1)], \partial_t F(\gamma_R(t), 1)\rangle dt \\ = &\int_0^{2\pi} \langle f[G(\gamma_R(t), 1)], \partial_t G(\gamma_R(t), 1)\rangle dt = \int_0^{2\pi} \langle f[G(\gamma_R(t), 0)], \partial_t G(\gamma_R(t), 0)\rangle dt. \end{aligned}$$

However, as $G(\cdot, 0) = -P(0)$ is constant and $F(\cdot, 0) = I$,

$$\begin{aligned} 0 &= \int_0^{2\pi} \langle f[G(\gamma_R(t), 0)], \partial_t G(\gamma_R(t), 0)\rangle dt \\ &= \int_0^{2\pi} \langle f[F(\gamma_R(t), 0)], \partial_t F(\gamma_R(t), 0)\rangle dt \\ &= \int_0^{2\pi} \langle f(\gamma_R(t)), \gamma_R'(t)\rangle\, dt = \int_0^{2\pi} (\sin^2 t + \cos^2 t)\, dt = 2\pi, \end{aligned}$$

a contradiction. □

A direct consequence is the **Brouwer FPT on a closed disc.**

Corollary 1. *Any continuous mapping $T : B_R \to B_R$ has a fixed point in B_R.*

3. A Proof of the Birkhoff–Kellogg Theorem on a Closed n-Ball Based on Differential n-Forms

The argument used in Section 2 for mappings in $\mathbb{R}^2$ can be extended to mappings in $\mathbb{R}^n$, using the basic properties of differential k-forms in $\mathbb{R}^n$. For $n = 2$, the differential 1-forms and differential $(n-1)$-forms coincide, and it is the last ones that are requested for extending the proof of Theorem 1 to arbitrary n. We leave to the motivated reader the work to write down this extension of the first approach and to realize that this generalization to dimension n of Lemma 1 is very cumbersome and lengthy. Fortunately a similar approach based on differential n-forms instead of $(n-1)$-forms has been introduced in [8], which, for $n = 2$, has the same length and technicality as the one used in Section 2, but keeps its simplicity for arbitrary n. We describe it in this section.

For $D \subset \mathbb{R}^n$ open, bounded and nonempty, we need the concept of differential $(n-1)$-forms and n-forms and suppose that the reader is familiar with the notions, notations and properties of differential k-forms $(1 \leq k \leq n)$ on D, wedge products, pull backs, exterior differentials and the Stokes–Cartan theorem for differential forms with compact support [11]. All the functions involved in differential forms are supposed to be of class C^2. We associate to the functions $f_j : \overline{D} \to \mathbb{R}$ $(j = 1, \ldots, n)$ the **differential 1-form** $\omega_f := \sum_{j=1}^n f_j\, dx_j$ **in** D, and the **differential** $(n-1)$**-form**

$$\nu_f = \sum_{j=1}^n (-1)^{j-1} f_j\, dx_1 \wedge \ldots \wedge \widehat{dx_j} \wedge \ldots \wedge dx_n,$$

where $\widehat{dx_j}$ means that the corresponding term is missing. We associate also to $g : \overline{D} \to \mathbb{R}^n$ the **differential** n**-form** $\mu_g = g\, dx_1 \wedge \ldots \wedge dx_n$. For example, given the function $w : D \to \mathbb{R}$ with partial derivatives $\partial_j w$, its **differential** $dw := \sum_{j=1}^n (\partial_j w)\, dx_j$ is the differential 1-form $\omega_{\nabla w}$.

Let $\Delta \subset \mathbb{R}^n$ be open, bounded and nonempty, $F : \Delta \times [0,1] \to D,\ (y, \lambda) \mapsto F(y, \lambda)$. For each fixed $\lambda \in [0,1]$,

$$\begin{aligned} F^*(\cdot,\lambda)\omega_f &= \sum_{j=1}^n [f_j \circ F(\cdot,\lambda)]\, dF_j(\cdot,\lambda) \\ &= \sum_{k=1}^n \left[\sum_{j=1}^n [f_j \circ F(\cdot,\lambda)] \partial_k F_j(\cdot,\lambda) \right] dy_k \quad (j = 1, \ldots, n) \end{aligned}$$

is well defined. To shorten the notations, we write F_j for $F_j(\cdot, \lambda)$. We define the **derivative with respect to** λ of $F^*\omega_f$ by

$$\partial_\lambda (F^*\omega_f) := \sum_{k=1}^n \partial_\lambda \left[\sum_{j=1}^n (f_j \circ F) \partial_k F_j \right] dy_k.$$

so that

$$\begin{aligned} \partial_\lambda (F^*\omega_f) &= \sum_{k=1}^n \sum_{j=1}^n [\partial_\lambda (f_j \circ F) \partial_k F_j + (f_j \circ F) \partial_\lambda \partial_k F_j]\, dy_k \\ &= \sum_{j=1}^n [\partial_\lambda (f_j \circ F)\, dF_j + (f_j \circ F)\, \partial_\lambda (dF_j)]. \end{aligned}$$

Furthermore,

$$\partial_\lambda (dF_j) = \sum_{k=1}^n (\partial_\lambda \partial_k F_j)\, dy_k = \sum_{k=1}^n (\partial_k \partial_\lambda F_j)\, dy_k = d(\partial_\lambda F_j) \quad (j = 1, \ldots, n).$$

On the other hand,

$$dF_1 \wedge \ldots \wedge dF_n = J_F\, dy_1 \wedge \ldots dy_n,$$

where $J_{F(\cdot,\lambda)}(y, \lambda)$ denotes the Jacobian of $F(\cdot, \lambda)$ at $(y, \lambda) \in \overline{\Delta} \times [0,1]$, and

$$\partial_\lambda [dF_1 \wedge \ldots \wedge dF_n] = \sum_{j=1}^n dF_1 \wedge \ldots \wedge \partial_\lambda dF_j \wedge \ldots \wedge dF_n.$$

The following two results replace Lemma 1 in Section 2. The first one shows that the differential n-form $\partial_\lambda (F^*\mu_g)$ is exact in Δ, i.e., is the exterior differential of a $(n-1)$-differential form in Δ.

Lemma 2. *For each* $\lambda \in [0,1]$, *we have*

$$\partial_\lambda(F^*\mu_g) = d\left[(g\circ F)\left(\sum_{j=1}^{n}(-1)^{j-1}\partial_\lambda F_j\, dF_1\wedge\ldots\wedge\widehat{dF_j}\wedge\ldots\wedge dF_n\right)\right].$$

Proof. We have

$$\begin{aligned}
\partial_\lambda(F^*\mu_g) &= \partial_\lambda(g\circ F)\, dF_1\wedge\ldots\wedge dF_n + (g\circ F)\,\partial_\lambda\,(dF_1\wedge\ldots\wedge dF_n)\\
&= \left(\sum_{j=1}^{n}(\partial_j g\circ F)\partial_\lambda F_j\right) dF_1\wedge\ldots\wedge dF_n\\
&+ (g\circ F)\left(\sum_{j=1}^{n} dF_1\wedge\ldots\wedge\partial_\lambda dF_j\wedge\ldots\wedge dF_n\right)\\
&= \sum_{j=1}^{n}(-1)^{j-1}(\partial_j g\circ F)\, dF_j\wedge\partial_\lambda F_j\, dF_1\wedge\ldots\wedge\widehat{dF_j}\wedge\ldots\wedge dF_n\\
&+ (g\circ F)\left(\sum_{j=1}^{n}(-1)^{j-1} d\,(\partial_\lambda F_j)\wedge dF_1\wedge\ldots\wedge\widehat{dF_j}\wedge\ldots\wedge dF_n\right)\\
&= \sum_{j=1}^{n}(-1)^{j-1}\left(\sum_{k=1}^{n}(\partial_k g\circ F)\, dF_k\right)\wedge\partial_\lambda F_j\, dF_1\wedge\ldots\wedge\widehat{dF_j}\wedge\ldots\wedge dF_n\\
&+ (g\circ F)\left(\sum_{j=1}^{n}(-1)^{j-1} d\left(\partial_\lambda F_j\, dF_1\wedge\ldots\wedge\widehat{dF_j}\wedge\ldots\wedge dF_n\right)\right)\\
&= d(g\circ F)\wedge\left(\sum_{j=1}^{n}(-1)^{j-1}\partial_\lambda F_j\, dF_1\wedge\ldots\wedge\widehat{dF_j}\wedge\ldots\wedge dF_n\right)\\
&+ (g\circ F)\, d\left(\sum_{j=1}^{n}(-1)^{j-1}\partial_\lambda F_j\, dF_1\wedge\ldots\wedge\widehat{dF_j}\wedge\ldots\wedge dF_n\right)\\
&= d\left[(g\circ F)\left(\sum_{j=1}^{n}(-1)^{j-1}\partial_\lambda F_j\, dF_1\wedge\ldots\wedge\widehat{dF_j}\wedge\ldots\wedge dF_n\right)\right] := d\nu_{g,F}.
\end{aligned}$$

□

Corollary 2. *If* $w \in C^2(\mathbb{R}^n,\mathbb{R})$, Δ *is open, bounded and* $F \in C^2(\overline{\Delta}\times[0,1],\mathbb{R}^n)$ *verify* $F(\partial\Delta\times[0,1]) \cap$ *supp* $w = \emptyset$, *then* $\int_\Delta F^*\mu_w$ *is independent of* λ *on* $[0,1]$.

Proof. Using Lemma 2, the assumption and Stokes–Cartan theorem, we get

$$\partial_\lambda\int_\Delta F^*\mu_w = \int_\Delta \partial_\lambda(F^*\mu_w) = \int_\Delta d\nu_{w,F} = \int_{\partial\Delta}\nu_{w,F} = 0.$$

□

Let $B_R := \{x\in\mathbb{R}^n : |x|\le R\}$ with $|x|$ the Euclidian norm. We now show that Proposition 2 allows a simple proof of the **Birkhoff–Kellogg FPT on a closed n-ball**, quite similar to that of Theorem 1.

Theorem 2. *Any continuous mapping* $T : B_R \to \mathbb{R}^n$ *such that* $T(\partial B_R) \subset B_R$ *has a fixed point in* B_R.

Proof. Assume that T has no fixed point in B_R. Then, $x - T(x) \neq 0$ for $x \in \partial B_R$, and for $(x,\lambda) \in \partial B_R\times[0,1)$, we have $|x-\lambda T(x)| \ge R - \lambda|T(x)| \ge (1-\lambda)R > 0$. Thus, $|x-\lambda T(x)| > 0$ for all $(x,\lambda)\in\partial B_R\times[0,1]$. On the other hand, for $(x,\lambda)\in\partial B_R\times[0,1]$, we have $\lambda x\in B_R$, $\lambda x - T(\lambda x)\neq 0$, and hence $|\lambda x - T(\lambda x)| > 0$ for all $(x, \lambda) \in \partial B(R) \times [0,1]$. By continuity, there exists $\delta > 0$ such that $|x - \lambda T(x)| > \delta$ for all $(x, \lambda) \in \partial B_R \times [0,1]$. Let $P : \mathbb{R}^n \to \mathbb{R}^n$ be a polynomial such that $\max_{B_R} |P - T| \leq \delta/2$, and define $F \in C^\infty(\mathbb{R}^n \times [0,1], \mathbb{R}^n)$ and $G \in C^\infty(\mathbb{R}^n \times [0,1], \mathbb{R}^n)$ by $F(x, \lambda) = \lambda x - P(\lambda x)$ and $G(x, \lambda) = x - \lambda P(x)$, so that $|F(x, \lambda)| \geq \delta/2$ and $|G(x, \lambda)| \geq \delta/2$ for all $(x, \lambda) \in \partial B_R \times [0,1]$. Let $w \in C^2(\mathbb{R}^n, \mathbb{R})$ with supp $w \subset B(\delta/2)$, the open ball of center 0 and radius $\delta/2$, and $\int_{B_R} w(y)\, dy = 1$. Then, by Proposition 2 with $\overline{\Delta} = B_R$, we get

$$0 = \int_{B_R} F^*(\cdot, 0)\mu_w = \int_{B_R} F^*(\cdot, 1)\mu_w = \int_{B_R} (I - P)^*\mu_W,$$

and

$$\begin{aligned} \int_{B_R} (I - P)^*\mu_w &= \int_{B_R} G^*(\cdot, 1)\mu_w = \int_{B_R} G^*(\cdot, 0)\mu_w = \int_{B_R} \mu_w \\ &= \int_{B_R} w(y)\, dy = 1, \end{aligned}$$

a contradiction. □

The **Brouwer FPT on a closed n-ball** is a special case.

Corollary 3. *Any continuous mapping $T : B_R \to B_R$ has a fixed point in B_R.*

4. Fixed Points, Homeomorphisms and Retractions in $\mathbb{R}^n$

Now, if $K \subset \mathbb{R}^n$, if there exists a homeomorphism $h : B^n \to K$, and if $T : K \to K$ is continuous, $h^{-1} \circ T \circ h : B^n \to B^n$ is continuous, has a fixed point x^* by Theorem 3, and $h(x^*) \in K$ is a fixed point of T. Consequently, we have a **Brouwer FPT for homeomorphic images of a closed n-ball.**

Theorem 3. *If $K \subset \mathbb{R}^n$ is homeomorphic to B_R, any continuous mapping $T : K \to K$ has a fixed point in K.*

For example, K can be any **closed n-interval** $[a_1, b_1] \times \ldots \times [a_n, b_n]$, or an **$n$-simplex** $\mathbb{R}^n_+ := \{x = \sum_{j=1}^n x_j e^j \in \mathbb{R}^n : x_j \geq 0, \sum_{j=1}^n x_j \leq 1\}$.

Remark 1. *In Theorem 3, the boundedness assumption on K cannot be omitted: a translation $x \mapsto x + a$ in $\mathbb{R}^n$ with $a \neq 0$ has no fixed point. The closedness assumption on K cannot be omitted as well: $T : (0,1) \to (0,1)$, $x \mapsto x^2$ has no fixed point in $(0,1)$. Theorem 3 does not hold for any closed bounded set: a nontrivial rotation of the closed annulus $A = \{x \in \mathbb{R}^2 : r_1 \leq |x| \leq r_2\}$ has no fixed point in A.*

We now introduce concepts and results due to Borsuk [12] which provide another class of sets on which the Brouwer FPT holds and simple proofs of various equivalent formulations of this theorem. We say that $U \subset V \subset \mathbb{R}^n$ is a **retract** of V if there exists a continuous mapping $r : V \to U$ such that $r = I$ on U (**retraction of V in U**). For example, B_R is a retract of $\mathbb{R}^n$, with a retraction r given by

$$r(x) = \begin{cases} x & \text{if} \quad |x| \leq R \\ R\frac{x}{|x|} & \text{if} \quad |x| > R. \end{cases} \tag{1}$$

Similarly, *for any $0 < R_1 \leq R_2$, B_{R_1} is a retract of B_{R_2}.*

Remark 2. *The Brouwer FPT on B_R implies the Birkhoff–Kellogg FPT on B_R. Indeed, if $T : B_R \to \mathbb{R}^n$ is continuous, $T(\partial B_R) \subset B_R$, and r is given by (1), then $r \circ T : B_R \to B_R$ is continuous and, by the Brouwer FPT 3, has a fixed point $x^* \in B_R$. If $|T(x^*)| > R$, $|x^*| = |r(T(x^*))| = R$ and $|T(x^*)| \leq R$, a contradiction. Thus, $|T(x^*)| \leq R$ and $x^* = T(x^*)$. Thus, the two statements are equivalent.*

Remark 3. *The Brouwer FPT has for immediate topological consequence the well-known* **no-retraction theorem**, *stating that* ∂B_R *is not a retract of* B_R *in* $\mathbb{R}^n$. *We do not repeat here the simple proof of this result and the proof of Brouwer FPT from the no-retraction theorem.*

An easy consequence of Theorem 3 is the **Leray–Schauder–Schaefer fixed point theorem**, a special case of a more general result obtained in 1934 by Leray and Schauder [13]. The proof given here is due to Schaefer [14].

Theorem 4. *Any continuous mapping* $T : B_R \subset \mathbb{R}^n \to \mathbb{R}^n$ *such that* $x \neq \lambda T(x)$ *for all* $(x, \lambda) \in \partial B_R \times (0,1)$ *has a fixed point in* B_R.

Proof. Let $r : \mathbb{R}^n \to B_R$ be the retraction of $\mathbb{R}^n$ onto B_R defined in Equation (1). Theorem 3 implies the existence of $x^* \in B_R$ such that $x^* = r(T(x^*))$. If $|T(x^*)| > R$, then $x^* = \frac{R}{|T(x^*)|}T(x^*)$, so that $|x^*| = R$ and $x^* = \lambda^* T(x^*)$ with $\lambda^* = \frac{R}{|T(x^*)|} < 1$, a contradiction with the assumption. Hence, $|T(x^*)| \leq R$ and $x^* = T(x^*)$. □

Remark 4. If $T : \partial B_R \to B_R$, it is clear that the assumption of Theorem 4 is satisfied. Thus *the Leray–Schauder–Schaefer FPT implies the Birkhoff–Kellogg FPT,* and hence *the two statements are quivalent.*

The Brouwer FPT holds for retracts of a closed ball.

Theorem 5. *If* $U \subset \mathbb{R}^n$ *is a retract of* B_R*, any continuous mapping* $T : U \to U$ *has a fixed point.*

Proof. Let $U = r(B_R)$ for some retraction $r : B_R \to U$. Then, $T \circ r : B_R \to U \subset B_R$ has a fixed point $x^* \in U$. Hence, $x^* = r(x^*)$, and $x^* = T(x^*)$. □

If $C \subset \mathbb{R}^n$ is non- empty, closed and convex, the **orthogonal projection** $p_C(x)$ **on** C of $x \in \mathbb{R}^n$, defined by $|p_C(x) - x| = \min_{y \in C} |y - x|$, is a retraction of $\mathbb{R}^n$ onto C [15]. Consequently, *C is a retract of any $B_R \supset C$*, giving a **Brouwer FPT on compact convex sets.**

Corollary 4. *If* $C \subset \mathbb{R}^n$ *is compact and convex, any continuous mapping* $T : C \to C$ *has a fixed point in* C.

5. Zeros of Continuous Mappings in $\mathbb{R}^n$

The first theorem on the existence of a zero for a mapping from B_R into $\mathbb{R}^n$ was first stated and proved for C^1 mappings by Bohl [16] in 1904, and extended to continuous mappings by Hadamard in 1910 [3], under the name **Poincaré–Bohl theorem**. It is a reformulation of the Leray–Schauder–Schaefer FPT Theorem 4.

Theorem 6. *Any continuous mapping* $f : B_R \to \mathbb{R}^n$ *such that* $f(x) \neq \mu x$ *for all* $x \in \partial B_R$ *and for all* $\mu < 0$ *has a zero in* B_R.

Proof. Define the continuous mapping $T : B_R \to \mathbb{R}^n$ by $T(x) = x - f(x)$. For $(x, \lambda) \in \partial B_R \times (0,1)$, we have, by assumption,

$$x - \lambda T(x) = (1 - \lambda)x + \lambda f(x) = \lambda \left[f(x) - \frac{\lambda - 1}{\lambda} x \right] \neq 0.$$

By Theorem 4, T has a fixed point x^* in B_R, which is a zero of f. □

In 1910, two years before the publication of [4], Hadamard, informed by a letter from Brouwer of the statement of his fixed point theorem, published a simple proof based on the Kronecker index (a forerunner of the Brouwer topological degree) in an appendix to an introductory analysis book of Tannery [3]. Hadamard's proof consisted in showing that Brouwer's assumption implies that the condition $\langle x, x - T(x)\rangle \geq 0$ holds for all $x \in \partial B_R$, where $\langle \cdot, \cdot \rangle$ denotes the usual inner product in $\mathbb{R}^n$. This condition implies the existence of a zero of $I - T$, because the assumption of the Poincaré–Bohl theorem 6 is satisfied. Hadamard's reasoning using the Kronecker index does not depend upon the special structure $I - T$ of the mapping in the inner product. Hence, it is natural (although not usual) to call **Hadamard theorem** the statement of existence of a zero for a continuous mapping $f : B_R \to \mathbb{R}^n$, when $x - T(x)$ is replaced by $f(x)$ in the inequality above, a statement which became in the year 1960 a key ingredient in the theory of monotone operators in reflexive Banach spaces. Using convex analysis, we give an extension to compact convex sets.

Let $C \subset \mathbb{R}^n$ be compact and convex and $p_C : \mathbb{R}^n \to C$ be the orthogonal projection of x on C [15]. Recall that $p_C(x)$ is characterized by the condition

$$\langle x - p_C(x), y - p_C(x)\rangle \leq 0 \quad \text{for all } y \in C. \tag{2}$$

For $x \in \partial C$, the set

$$N_x := \{v \in \mathbb{R}^n : \langle v, y - x\rangle \leq 0 \ \text{ for all } \ y \in C\}$$

is nonempty and called the **normal cone** to C at x, and its elements v are called the **outer normals** to C at x. The relation in Equation (2) shows that, for each $x \notin C$, $x - p(x) \in N_{p(x)} \setminus \{0\}$. It can also be shown that each $x \in \partial C$ is the orthogonal projection of some $z \notin C$, so that $N_x = \{z \in \mathbb{R}^n \setminus C : p(z) = x\}$. The **Hadamard theorem on a convex compact set** follows in a similar way as Theorem 6 from the Brouwer FPT 3.

Theorem 7. *If $C \subset \mathbb{R}^n$ is a compact and convex, any continuous $f : C \to \mathbb{R}^n$ such that $\langle v, f(x)\rangle \geq 0$ for all $x \in \partial C$ and all $v \in N_x$ has a zero in C.*

Proof. Let $T : \mathbb{R}^n \to \mathbb{R}^n$ be defined by $T = p_C - f \circ p_C$. Then, for all $x \in \mathbb{R}^n$,

$$|T(x)| \leq |p_C(x)| + |f(p_C(x))| \leq \max_{x \in C} |x| + \max_{y \in C} |f(y)| := R,$$

and T maps B_R into itself. By Theorem 3, there exists $x^* \in B_R$ such that $x^* = p_C(x^*) - f(p_C(x^*))$. If $x^* \notin C$, the assumption implies that

$$0 < |x^* - p_C(x^*)|^2 = -\langle x^* - p_C(x^*), f(p_C(x^*))\rangle \leq 0,$$

a contradiction. Thus, $x^* \in C$, $x^* = p_C(x^*)$ and $f(x^*) = 0$. □

Corollary 5. *Any continuous mapping $f : B_R \to \mathbb{R}^n$ such that $\langle x, f(x)\rangle \geq 0$ for all $x \in \partial B_R$ has a zero in B_R.*

Proof. For each $x \in \partial B_R$, $N_x = \{\lambda x : \lambda > 0\}$, and we apply Theorem 7. □

Remark 5. *As shown when mentioning Hadamard's contribution, Theorem 5 implies the Brouwer FPT, and even the Birkhoff–Kellopg FPT, on B_R. Consequently, those statements are equivalent.*

Some twenty years before the publication of Brouwer's paper [4], Poincaré [17] stated in 1883 a theorem about the existence of a zero of a continuous mapping $f : P = [-R_1, R_1] \times \cdots \times [-R_n, R_n] \to \mathbb{R}^n$ when, for each $i = 1, \ldots, n$, f_i takes opposite signs on the opposite faces of P

$$P_i^- := \{x \in P : x_i = -R_i\}, \quad P_i^+ := \{x \in P : x_i = R_i\} \quad (i = 1, \ldots, n).$$

Poincaré's proof just told that the result was a consequence of the Kronecker index, which is correct but sketchy. The statement, forgotten for a while, was rediscovered by Cinquini [18] in 1940 with an inconclusive proof, and shown to be equivalent to the Brouwer FPT on P one year later by Miranda [19]. Many other proofs have been given since, and we again refer to [7,20] for a more complete history, variations and references, and to [21–23] for useful generalizations to more complicated sets than closed n-intervals. Here, we obtain the **Poincaré–Miranda theorem on a closed n-interval** as a special case of Theorem 7.

Corollary 6. *Any continuous mapping $f : P \to \mathbb{R}^n$ such that $f_i(x) \leq 0$ for all $x \in P_i^-$ and $f_i(x) \geq 0$ for all $x \in P_i^+$ $(i = 1, \ldots, n)$ has a zero in P.*

Proof. If x is in the (relative) interior of the face P_i^-, then $N_x = \{-\lambda e_i : \lambda > 0\}$, where $(e_1, e_2, \ldots, e_n)$ is the orthonormal basis in $\mathbb{R}^n$, and the assumption of Theorem 7 becomes $-f_i(x) \geq 0$, i.e., $f_i(x) \leq 0$. Similarly, if x is in the (relative) interior to the face P_i^+, then $N_x = \{\lambda e_i : \lambda > 0\}$, and the assumption of Theorem 7 becomes $f_i(x) \geq 0$. Of course, $-\lambda e_i$ and λe_i $(\lambda > 0)$ also belong to the respective normal cones for $x \in P_i^-$ and P_i^+ respectively, and if, say, $x \in P_i^- \cap P_j^+$ then $v = -\lambda e_i + \mu e_j \in N_x$ for all $\lambda, \mu > 0$, and $\langle v, f(x)\rangle = -\lambda f_i(x) + \mu f_k(x) \geq 0$. In general, when x belongs to the intersection of several faces of P, N_x will be made of the linear combination of the e_i corresponding to the indices of the faces, with a negative coefficient for a face having symbol $-$ and positive coefficient for a face having symbol $+$, so that, using the assumption, $\langle v, f(x)\rangle \geq 0$ for all $x \in \partial P$ and all $v \in N_x$. The result follows from Theorem 7. □

Remark 6. *Corollary 6 implies the Brouwer FPT on P. Indeed, if $T : P \to P$ is continuous, and if we set $f = I - T$, then, as $-R_i \leq T_i(x) \leq R_i$ for all $x \in \partial P$, we have, for $x \in P$ such that $x_i = -R_i$, $f_i(x) = x_i - T_i(x) = -R_i - T_i(x) \leq 0$, and, for $x \in P$ such that $x_i = R_i$, $f_i(x) = x_i - T_i(x) = R_i - T_i(x) \geq 0$. Thus f has at least one zero in P, which is a fixed point of T. Consequently, the two statements are equivalent.*

Remark 7. *Both the Hadamard theorem on B_R and the Poincaré–Miranda theorem can be seen as distinct n-dimensional generalizations of the Bolzano theorem to closed ball and n-intervals respectively.*

Remark 8. *Using the Brouwer degree, it is easy to obtain the conclusion of the Hadamard Theorem 7 for a compact convex neighborhood of 0 under the weaker condition that for each $x \in \partial C$, there exists $v \in N_x$ such that $\langle v, f(x)\rangle \geq 0$. No proof based only upon the Brouwer FPT seems to be known.*

If $C \subset \mathbb{R}^n$ is a compact convex set and $g : C \to \mathbb{R}$ is of class C^1, then g reaches its minimum on C at some $x^* \in C$ for which

$$g(x^* + \lambda(v - x^*)) - g(x^*) \geq 0 \text{ for all } v \in C \text{ and for all } \lambda \in [0, 1],$$

so that, dividing both members by λ and letting $\lambda \to 0+$, we obtain $\langle \nabla g(x^*), v - x^*\rangle \geq 0$ for all $v \in C$, where ∇g denotes the gradient of g. For example, if $u \in \mathbb{R}^n$ is fixed and $g : C \to \mathbb{R}$ is defined by $g(x) = (1/2)|x - u|^2$, the minimization problem corresponds to the definition of $p_C(u)$, and, as $\nabla g(x) = x - u$, the inequality above is just Equation (2). In 1966, Hartman and Stampacchia [24] proved that the existence of such a x^* still holds when ∇g is replaced by an arbitrary continuous function $f : C \to \mathbb{R}^n$. When C is a simplex, the same result was proved independently the same year by Karamardian [25]. We give here a proof, due to Brezis (see [26]) and based upon Brouwer's FPT, of the **Hartman–Stampacchia theorem on variational inequalities**.

Theorem 8. *If $C \subset \mathbb{R}^n$ is compact, convex and $f : C \to \mathbb{R}^n$ continuous, there exists $x^* \in C$ such that $\langle f(x^*), v - x^*\rangle \geq 0$ for all $v \in C$.*

Proof. The Brouwer FPT on C (Corollary 4) applied to the continuous mapping $p_C \circ (I - f) : C \to C$ implies the existence of $x^* \in C$ such that

$$x^* = p_C(x^* - f(x^*)). \tag{3}$$

Taking $x = x^* - f(x^*)$ in Equation (2) and using Equation (3), one gets

$$\langle x^* - f(x^*) - x^*, v - x^* \rangle \leq 0 \text{ for all } v \in C,$$

which is the requested inequality. □

Remark 9. *The conclusion of Theorem 8 is called a* **variational inequality**. *In the terminology of the theory of convex sets [15], the conclusion of Theorem 8 means that there exists $x^* \in C$ such that either $f(x^*) = 0$ or $f(x^*) \neq 0$ and $H := \{y \in \mathbb{R}^n : \langle f(x^*), y - x^* \rangle = 0\}$ is a* **supporting hyperplane** *for C passing through x^* i.e., C is entirely contained in one of the two closed half-spaces determined by H.*

Remark 10. *The Brouwer FPT on C (Corollary 4) also follows from the Hartman–Stampacchia theorem. Indeed, if $T : C \to C$ is continuous and x^* is given by Theorem 8 applied to $f = I - T$, then, taking $v = T(x^*) \in C$ in the variational inequality, we obtain $0 \leq \langle x^* - T(x^*), T(x^*) - x^* \rangle = -|x^* - T(x^*)|^2 \leq 0$, so that $x^* = T(x^*)$. Hence, the two statements are equivalent.*

Remark 11. *If x^* and $x^\#$ are two distinct solutions of the variational inequality, then*

$$\langle f(x^*), x^\# - x^* \rangle \geq 0, \quad \langle f(x^\#), x^* - x^\# \rangle \geq 0,$$

and hence $\langle f(x^*) - f(x^\#), x^* - x^\# \rangle \leq 0$. Consequently, the variational inequality has a unique solution if f satisfies the condition $\langle f(x) - f(y), x - y \rangle > 0$ for all $x \neq y \in C$, i.e., if f is **strictly monotone** on C.

6. A Direct Approach for Holomorphic Functions in $\mathbb{C}$

The assumption of the Bolzano theorem for a continuous function $f : [-R, R] \to \mathbb{R}$ can be, without loss of generality, be written $f(-R) \leq 0 \leq f(R)$ or, equivalently, $xf(x) \geq 0$ for $|x| = R$. In 1982, Shih [27] proposed a version of the Bolzano theorem for a complex function f holomorphic on a suitable bounded open neighborhood $\Omega \subset \mathbb{C}$ of 0 and continuous on $\overline{\Omega}$. He showed that f has a unique zero in Ω when $\Re[\overline{z}f(z)] > 0$ on $\partial\Omega$, using the Rouché theorem applied to $f(z)$ and $g(z) = \alpha z$ for a suitable real α. As $\Re[\overline{z}f(z)] = \Re z \cdot \Re f(z) + \Im z \cdot \Im f(z)$, Shih's condition is just Hadamard's one in Theorem 5 with strict inequality sign. Following the approach introduced in [10], we show in this section that, when the (non strict) Hadamard condition holds on the boundary of a ball, the existence of a zero of a holomorphic function results in a very simple way from an immediate consequence of the Cauchy integral theorem. The same is true for a Poincaré–Miranda theorem on a rectangle, giving another extension of the Bolzano theorem to complex functions. The Brouwer's FPT for holomorphic functions on a closed ball or a closed rectangle follow immediately.

We suppose the reader familiar with the concepts of **holomorphic function** f, **piecewise** C^k **cycle** γ, and **integral** $\int_\gamma f(z)\,dz$ **of** f **along** γ [28]. We denote by $B(R)$ the open disc of center 0 and radius $R > 0$ in $\mathbb{C}$, and by B_R the corresponding closed disc. Let $\gamma_R : [0, 2\pi] \to \partial B_R$, $t \mapsto Re^{it}$ be the standard C^∞-cycle whose image is ∂B_R. The **Cauchy integral theorem on a circle** is proved here in a simple way, reminiscent of Cauchy's proof in 1825 [29], reworked by Falk in 1883 [30], and similar in spirit to the proof of Lemma 2.

Proposition 1. *If $f : B_R \to \mathbb{C}$ is continuous on B_R and holomorphic on $B(R)$, then $\int_{\gamma_R} f(z)\,dz = 0$.*

Proof. Define $\Gamma : [0,1] \times [0,2\pi] \to B_R$ by $\Gamma(\lambda, t) = \lambda\gamma_R(t)$, so that $\Gamma(1,\cdot) = \gamma_R$ and $\Gamma(0,\cdot)$ is the constant zero mapping. To show that $\lambda \mapsto \int_{\Gamma(\lambda,\cdot)} f(z)\,dz$ is constant in $(0,1)$, we have (with differentiation under the integral sign easily justified and $'$ denoting the derivative with respect to z)

$$\begin{aligned}
\partial_\lambda \int_{\Gamma(\lambda,\cdot)} f(z)\,dz &= \partial_\lambda \int_0^{2\pi} f(\Gamma(\lambda,t))\partial_t\Gamma(\lambda,t)\,dt \\
&= \int_0^{2\pi} [f'(\Gamma(\lambda,t))\partial_\lambda\Gamma(s,t)\partial_t\Gamma(\lambda,t) + f(\Gamma(\lambda,t))\partial_\lambda\partial_t\Gamma(\lambda,t)]\,dt \\
&= \int_0^{2\pi} [\partial_t\{f(\Gamma(\lambda,t))\}\partial_\lambda\Gamma(\lambda,t) + f(\Gamma(\lambda,t))\partial_t\partial_\lambda\Gamma(\lambda,t)]\,dt \\
&= \int_0^{2\pi} \partial_t[f(\Gamma(\lambda,t))\partial_\lambda\Gamma(\lambda,t)]\,dt \\
&= f(\Gamma(\lambda,2\pi))\partial_\lambda\Gamma(\lambda,2\pi) - f(\Gamma(\lambda,0))\partial_\lambda\Gamma(\lambda,0) = 0.
\end{aligned}$$

By continuity, $\lambda \mapsto \int_{\Gamma(\lambda,\cdot)} f(z)\,dz$ is constant in $[0,1]$, and hence

$$\int_{\gamma_R} f(z)\,dz = \int_{\Gamma(1,\cdot)} f(z)\,dz = \int_{\Gamma(0,\cdot)} f(z)\,dz = 0.$$

□

Let $a > 0$, $b > 0$, $P = \{z \in \mathbb{C} : -a \le \Re z \le a,\ -b \le \Im z \le b\}$ be the corresponding closed rectangle in $\mathbb{C}$, and let us introduce the continuous mapping $\rho : [0,4] \to \partial P$ of class C^∞ on $(0,1) \cup (1,2) \cup (2,3) \cup (3,4)$ defined by

$$\rho(t) = \begin{cases} -a + 2ta - ib & \text{if } t \in [0,1] \\ a + i[-b + 2(t-1)b] & \text{if } t \in [1,2] \\ a - 2(t-2)a + ib & \text{if } t \in [2,3] \\ -a + i[b - 2(t-3)b] & \text{if } t \in [3,4], \end{cases} \tag{4}$$

whose image $\rho([0,4]) = \partial P$. We state and prove the **Cauchy's integral theorem on the boundary of a rectangle.**

Proposition 2. *If $f : P \to \mathbb{C}$ is continuous on P and holomorphic on int P, then $\int_\rho f(z)\,dz = 0$.*

Proof. It is entirely similar to that of Proposition 1. If we define $R : [0,1] \times [0,4] \to P$ by $R(\lambda,t) = \lambda\rho(t)$, the integral $\int_{R(\lambda,\cdot)} f(z)\,dz$ has to be decomposed into four integrals over $[0,1]$, $[1,2]$, $[2,3]$, and $[3,4]$, respectively, of $f(R(\lambda,t))\partial_t R(\lambda,t)$, and each integral has to be differentiated with respect to λ separately. The details are left to the reader. □

Propositions 1 and 2 immediately imply the following simple **theorem for the existence of a zero of f.**

Proposition 3. *Any function $f : B_R \to \mathbb{C}$ (respectively, $f : P \to \mathbb{C}$) holomorphic on $B(R)$ (respectively, int P), continuous on B_R (respectively, P), different from zero on ∂B_R (respectively, ∂P) and such that*

$$\int_{\gamma_R} \frac{dz}{f(z)} \neq 0 \quad \left(\text{resp. } \int_\rho \frac{dz}{f(z)} \neq 0\right)$$

has a zero in $B(R)$ (respectively, P).

Proof. It is entirely similar in both cases and we prove it for B_R. If f has no zero in $B(R)$, then $z \mapsto \frac{1}{f(z)}$ is holomorphic on $B(R)$ and continuous on B_R. By Proposition 1, $\int_{\gamma_R} \frac{dz}{f(z)} = 0$, a contradiction to the assumption. □

Proposition 3 provides a very simple proof of the **Hadamard theorem for a holomorphic function on B_R**.

Theorem 9. *Any function $f : B_R \to \mathbb{C}$ holomorphic on $B(R)$, continuous on B_R and such that $\Re[\overline{z}f(z)] \geq 0$ for all $z \in \partial B_R$, has a zero in B_R.*

Proof. For each integer $k \geq 1$, define $f_k : B_R \to \mathbb{C}$ by $f_k(z) = k^{-1}z + f(z)$. Each f_k has the regularity properties of f and is such that, for any $z \in \partial B_R$, $\Re[\overline{z}f_k(z)] = k^{-1}R^2 + \Re[\overline{z}f(z)] > 0$, so that $f_k(z) \neq 0$ for all $z \in \partial B_R$, and

$$\begin{aligned}
\Im\left[\int_{\gamma_R} \frac{dz}{f_k(z)}\right] &= \Im\left[\int_{\gamma_R} \frac{z\overline{z}}{\overline{z}f_k(z)}\frac{dz}{z}\right] \\
&= \Im\left[\int_{\gamma_R} \frac{|z|^2\{\Re[\overline{z}f_k(z)] - i\Im[\overline{z}f_k(z)]\}}{|\overline{z}f_k(z)|^2}\frac{dz}{z}\right] \\
&= \Im\left[\int_0^{2\pi} \frac{i\Re[Re^{-it}f_k(Re^{it})] + \Im[Re^{-it}f(Re^{it})]}{|f_k(Re^{it})|^2} dt\right] \\
&= \int_0^{2\pi} \frac{\Re[Re^{-it}f_k(Re^{it})]}{|f_k(Re^{it})|^2}\, dt > 0.
\end{aligned}$$

By Proposition 3, for each $k \geq 1$, f_k has a zero z_k in $B(R)$, and, by the Bolzano–Weierstrass theorem, a subsequence $(z_{k_n})_{n\geq 1}$ of $(z_k)_{k\geq 1}$ converges to some $z^* \in B_R$ such that $0 = \lim_{n\to\infty}[k_n^{-1}z_{k_n} + f(z_{k_n})] = f(z^*)$. □

The **Birkhoff–Kellog FPT for a holomorphic function on a disc** is a direct consequence of Theorem 9.

Corollary 7. *Any function $T : B_R \to \mathbb{C}$ continuous on B_R, holomorphic on $B(R)$ and such that $T(\partial B_R) \subset B_R$ has a fixed point in B_R.*

Proof. For each $z \in \partial B_R$, one has $\Re\{\overline{z}[z - T(z)]\} \geq R^2 - |z||T(z)| \geq 0$. □

Example 1. *For any integer $m \geq 1$, the mapping T defined by $T(z) = \frac{z}{2}(z^m + 1)$ is such that for $|z| = 1$, $|T(z)| \leq \frac{|z|}{2}(|z|^m + 1) \leq 1$. There is no uniqueness as T has the fixed points 0 and 1 in B_1.*

Let $P_1^- = \{-a + iy : y \in [-b, b]\}$, $P^-1+ = \{a + iy : y \in [-b, b]\}$, $P_2^- = \{x - ib : x \in [-a, a]\}$ and $P_2^+ = \{x + ib : x \in [-a, a]\}$ be the opposite vertical and horizontal sides of P, respectively. Proposition 3 provides a **Poincaré–Miranda theorem for a holomorphic function on a rectangle**.

Theorem 10. *Any function $f : P \to \mathbb{C}$ continuous on P, holomorphic on int P and such that $\Re f(z) \leq 0$ for all $z \in P_1^-$, $\Re f(z) \geq 0$ for all $z \in P_1^+$, $\Im f(z) \leq 0$ for all $z \in P_2^-$ and $\Im f(z) \geq 0$ for all $z \in P_2^+$ has a zero in P.*

Proof. For each integer $k \geq 1$, the function f_k defined on P by $f_k(z) = k^{-1}z + f(z)$ is such that $\Re f_k(z) < 0$ for $z \in P_1^-$, $\Re f_k(z) > 0$ for $z \in P_1^+$, $\Im f_k(z) < 0$ for $z \in P_2^-$, and $\Im f_k(z) < 0$ for

$z \in P_2^+$. Hence, $f_k(z) \neq 0$ for each $z \in \partial P$. Let $\rho : [0,4] \to \Omega$ be the cycle defined by Equation (4). By the assumptions,

$$
\begin{aligned}
& \Im\left[\int_\rho \frac{dz}{f_k(z)}\right] = \Im\left\{\int_\rho |f_k(z)|^{-2}[\Re f_k(z) - i\Im f_k(z)]\,dz\right\} \\
= & \int_\rho |f_k(z)|^{-2}[-\Im f_k(z)\,dx + \Re f_k(z)\,dy] \\
= & -\int_0^1 |f_k(\rho(t))|^{-2}\Im f_k[\rho(t)]2a\,dt + \int_1^2 |f_k(\rho(t))|^{-2}\Re f_k[\rho(t)]2b\,dt \\
- & \int_2^3 |f_k(\rho(t))|^{-2}\Im f_k[\rho(t)]2a\,dt + \int_3^4 |f_k(\rho(t))|^{-2}\Re f_k[\rho(t)]2b\,dt \\
= & -\int_{-a}^a |f_k(s-ib)|^{-2}\Im f_k(s-ib)\,ds + \int_{-b}^b |f_k(a+it)|^{-2}\Re f_k(a+it)\,ds \\
+ & \int_{-a}^a |f_k(s+ib)|^{-2}\Im f_k(s+ib)\,dt - \int_{-b}^b |f_k(-a+is)|^{-2}\Re f_k(-a+is)\,ds \\
= & \int_{-a}^a \left[-|f_k(s-ib)|^{-2}\Im f_k(s-ib) + |f_k(s+ib)|^2\Im f_k(s+ib)\right]\,ds \\
+ & \int_{-b}^b \left[|f_k(a+is)|^{-2}\Re f_k(a+is) - |f_k(-a+is)|^{-2}\Re f_k(-a+is)\right]\,ds > 0,
\end{aligned}
$$

For $k \geq 1$, Proposition 3 implies the existence of $z_k \in \operatorname{int} P$ such that $k^{-1}z_k + f(z_k) = 0$. Using the Bolzano–Weierstrass theorem, a subsequence $(z_{k_n})_{n\geq 1}$ converges to some $z^* \in P$ such that $0 = \lim_{n\to\infty}[k_n^{-1}z_{k_n} + f(z_{k_n})] = f(z^*)$. □

Example 2. *Let the holomorphic function* $f : \mathbb{C} \to \mathbb{C}$ *be defined by* $f(z) = z^3 + 4z + 1 + i$. *Taking* $P = \{z \in \mathbb{C} : \Re z \in [-1,1]$ *and* $\Im z \in [-1,1]\}$, *one has*

$$
\begin{aligned}
z \in P_1^- &\Rightarrow \Re f(z) = -4 + 3y^2 < 0, \quad z \in P_1^+ \Rightarrow \Re f(z) = 6 - 3y^2 > 0 \\
z \in P_2^- &\Rightarrow \Im f(z) = -3x^2 - 2 < 0, \quad z \in P_2^+ \Rightarrow \Im f(z) = 3x^2 + 4 > 0,
\end{aligned}
$$

and f *has a zero in* $[-1,1] \times [-1,1]$.

A direct consequence of Theorem 10 is the **Birkhoff–Kellogg FPT for a holomorphic function on a rectangle**.

Corollary 8. *Any function* $T : P \to \mathbb{C}$ *continuous on* P, *holomorphic on int* P, *and such that* $T(\partial P) \subset P$ *has a fixed point in* P.

Proof. Define $f : P \to \mathbb{C}$ by $f(z) = z - T(z)$ for all $z \in P$. The assumption $T(\partial P) \subset P$ is equivalent to $-a \leq \Re T(z) \leq a$ and $-b \leq \Im T(z) \leq b$ for all $z \in \partial P$, and, hence, if $z \in P_1^-$, $\Re f(z) = -a - \Re T(z) \leq 0$, if $z \in P_1^+$, $\Re f(z) = a - \Re T(z) \geq 0$, if $z \in P_2^-$, $\Im f(z) = -b - \Im T(z) \leq 0$, and if $z \in P_2^+$, $\Im f(z) = b - \Im T(z) \geq 0$. Thus, by Theorem 10, f has a zero in P and T a fixed point in P. □

Funding: This research received no external funding.

Conflicts of Interest: The author declares no conflict of interest.

References

1. Bolzano, B. *Rein analytisches Beweis des Lehrsatzes dass zwischen je zwey Werthen, die ein entgegengesetztes Resultat gewähren, wenigsten eine reelle Wurzel der Gleichung liege*; Abh. K. Gesells. Wiss.: Prag, Czech, 1817.
2. Cauchy, A. *Cours d'analyse de l'École Royale Polytechnique*; Ire Partie Analyse Algébrique: Debure, Paris, 1821.

3. Hadamard, J. *Sur quelques applications de l'indice de Kronecker, Note additionnelle*, 2nd ed.; 'Introduction à la théorie des Fonctions d'une Variable' de J. Tannery; Hermann: Paris, France, 1910; Volume 2, pp. 437–477.
4. Brouwer, L.E.J. Ueber Abbildungen von Mannigfaltigkeiten. *Math. Ann.* **1912**, *71*, 97–115. [CrossRef]
5. Birkhoff, G.D.; Kellogg, O.D. Invariant points in function space. *Trans. Amer. Math. Soc.* **1992**, *23*, 96–115. [CrossRef]
6. Rothe, E. Zur Theorie der topologischen Ordnung und der Vektorfelder in Banachschen Räumen. *Compos. Math.* **1937**, *5*, 177–197.
7. Mawhin, J. Le théorème du point fixe de Brouwer: Un siècle de métamorphoses. *Sci. Tech. Perspect.* **2007**, *10*, 175–220.
8. Mawhin, J. Simple proofs of various fixed point and existence theorems based on exterior calculus. *Math. Nachr.* **2005**, *278*, 1607–1614. [CrossRef]
9. Mawhin, J. Simple proofs of the Hadamard and Poincaré–Miranda theorems using the Brouwer fixed point theorem. *Am. Math. Mon.* **2019**, *126*, 260–263. [CrossRef]
10. Mawhin, J. Bolzano's theorems for holomorphic mappings. *Chin. Ann. Math.* **2017**, *38*, 563–578. [CrossRef]
11. Flanders, H. *Differential Forms with Applications to the Physical Sciences*; Academic Press: New York, NY, USA, Reprinted Dover: New York, NY, USA, 1989.
12. Borsuk, K. Sur les rétractes. *Fundam. Math.* **1931**, *17*, 152–170. [CrossRef]
13. Leray, J.; Schauder, J. Topologie et équations fonctionnelles. *Ann. Scient. École Norm. Sup.* **1934**, *51*, 45–78. [CrossRef]
14. Schaefer, H. Ueber die Methode der a priori Schranken. *Math. Ann.* **1955**, *129*, 415–416. [CrossRef]
15. Hiriart-Urruty, J.B.; Lemaréchal, C. *Convex Analysis and Minimization Algorithms I*; Springer: Berlin, Germany, 1993.
16. Bohl, P. Ueber die Bewegung eines mechanischen Systems in der Nähe einer Gleichgewichtslage. *J. Reine Angew. Math.* **1904**, *127*, 179–276.
17. Poincaré, H. Sur certaines solutions particulières du problème des trois corps. *C R. Acad. Sci. Paris* **1883**, *97*, 251–252.
18. Cinquini, S. Problemi di valori al contorno per equazioni differenziali di ordine *n*. *Ann. Della Sc. Norm. Super.-Pisa Sci.* **1940**, *9*, 61–77.
19. Miranda, C. Un' osservazione su un teorema di Brouwer. *Cons. Naz. Delle Ric.* **1940**, *3*, 5–7.
20. Mawhin, J. Variations on Poincaré-Miranda's theorem. *Adv. Nonlinear Stud.* **2013**, *13*, 209–217. [CrossRef]
21. Fonda, A.; Gidoni, P. Generalizing the Poincaré-Miranda theorem: The avoiding cones condition. *Ann. Mat. Pura Appl.* **2016**, *195*, 1347–1371. [CrossRef]
22. Pireddu, M.; Zanolin, F. Fixed points for dissipative-repulsive systems and topological dynamics of mappings defined on *N*-dimensional cells. *Adv. Nonlinear Stud.* **2005**, *5*, 411–440. [CrossRef]
23. Pireddu, M.; Zanolin, F. Cutting surfaces and applications to periodic points and chaotic-like dynamics. *Topol. Methods Nonlinear Anal.* **2007**, *30*, 279–319.
24. Hartman, P.; Stampacchia, G. On some non-linear elliptic differential-functional equations. *Acta Math.* **1966**, *115*, 271–310. [CrossRef]
25. Karamardian, S. *Duality in Mathematical Programming*; ORC 66-2; Operations Research Center, University California: Berkeley, CA, USA, 1966.
26. Kinderlehrer, D.; Stampacchia, G. *An Introduction to Variational Inequalities and Their Applications*; Academic Press: New York, NY, USA, 1980.
27. Shih, M.H. An analog of Bolzano's theorem for functions of a complex variable. *Am. Math. Mon.* **1982**, *89*, 210–211. [CrossRef]
28. Ahlfors, L. *Complex Analysis*, 2nd ed.; Academic Press: New York, NY, USA, 1966.
29. Cauchy, A. Mémoire sur les intégrales définies. *Mém. Acad. Sci. Paris* **1827**, *15*, 41–89.
30. Falk, M. Extrait d'une lettre adressée à M. Hermite. *Bull. Sci. Math. Astron.* **1883**, *7*, 137–139.

mathematics

MDPI

Article

A Sub-Supersolution Approach for Robin Boundary Value Problems with Full Gradient Dependence

Dumitru Motreanu [1,*], Angela Sciammetta [2] and Elisabetta Tornatore [2]

1 Department of Mathematics, University of Perpignan, 66860 Perpignan, France
2 Department of Mathematics and Computer Science, University of Palermo, 90123 Palermo, Italy; angela.sciammetta@unipa.it (A.S.); elisa.tornatore@unipa.it (E.T.)
* Correspondence: motreanu@univ-perp.fr

Received: 19 March 2020; Accepted: 19 April 2020; Published: 27 April 2020

Abstract: The paper investigates a nonlinear elliptic problem with a Robin boundary condition, which exhibits a convection term with full dependence on the solution and its gradient. A sub-supersolution approach is developed for this type of problems. The main result establishes the existence of a solution enclosed in the ordered interval formed by a sub-supersolution. The result is applied to find positive solutions.

Keywords: nonlinear elliptic problem; Robin boundary condition; gradient dependence; sub-supersolution; positive solution

1. Introduction

In this paper we study the following nonlinear elliptic boundary value problem

$$\begin{cases} -\mathrm{div}(A(x,\nabla u)) + \alpha(x)|u|^{p-2}u = f(x,u,\nabla u) & \text{in } \Omega \\ A(x,\nabla u)\cdot\nu(x) + \beta(x)|u|^{p-2}u = 0 & \text{on } \partial\Omega \end{cases} \tag{1}$$

on a bounded domain $\Omega \subset \mathbb{R}^N$ with $N \geq 3$ and with a boundary $\partial\Omega$ of class C^1. The notation $\nu(x)$ stands for the unit exterior normal at any $x \in \partial\Omega$ and p is a real number with $1 < p < +\infty$. We note that, in the stated problem, the boundary condition is of Robin type.

We describe the data entering our problem. The leading differential part of the equation in (1) is the term $\mathrm{div}(A(x,\nabla u))$ in divergence form driven by the map $A : \overline{\Omega} \times \mathbb{R}^N \to \mathbb{R}^N$ which is composed with the (weak) gradient ∇u of the solution $u : \Omega \to \mathbb{R}$. No homogeneity condition is required for the map A. Precisely, we assume that $A : \overline{\Omega} \times \mathbb{R}^N \to \mathbb{R}^N$ is continuous and fulfills the conditions:

(A1) There exist constants c_1 and c_2 with $0 < c_1 \leq c_2$ such that

$$A(x,\xi)\cdot\xi \geq c_1|\xi|^p \text{ and } |A(x,\xi)| \leq c_2(|\xi|^{p-1}+1) \text{ for all } (x,\xi) \in \overline{\Omega}\times\mathbb{R}^N.$$

(A2) For all $x \in \Omega$, $A(x,\xi)$ is strictly monotone in ξ.

Here and subsequently we denote by $|\cdot|$ and $\cdot$ the standard Euclidean norm and scalar product on $\mathbb{R}^N$, respectively.

As important examples of operators $\mathrm{div}(A(x,\nabla u))$ complying with the preceding hypotheses we mention: the p-Laplacian $\Delta_p u := \mathrm{div}(|\nabla u|^{p-2}\nabla u)$ where $A(x,\xi) = |\xi|^{p-2}\xi$, the (p,q)-Laplacian $\Delta_p u + \Delta_q u := \mathrm{div}((|\nabla u|^{p-2} + |\nabla u|^{q-2})\nabla u)$ where $1 < q < p < +\infty$ and $A(x,\xi) = |\xi|^{p-2}\xi + |\xi|^{q-2}\xi$, the generalized p-mean curvature operator $\mathrm{div}((1+|\nabla u|^2)^{\frac{p-2}{2}}\nabla u)$ where $A(x,\xi) = (1+|\xi|^2)^{\frac{p-2}{2}}$ as well as numerous weighted versions.

The values of u on $\partial\Omega$ in the boundary condition of (1) are in the trace sense, whereas $A(x, \nabla u) \cdot \nu(x)$ represents the co-normal derivative of u associated with A. For more details we refer to ([1], pages 7–9) and ([2], Section 2). In the statement of problem (1) we fix the functions $\alpha \in L^\infty(\Omega)$ and $\beta \in L^\infty(\partial\Omega)$ satisfying $\alpha(x) \geq 0$ for almost everywhere (in short a.e.) $x \in \Omega$ and $\beta(x) \geq 0$ for a.e. $x \in \partial\Omega$, $\beta \not\equiv 0$, where $\partial\Omega$ is endowed with the $(N-1)$-dimensional Hausdorff measure. Contrary to the Neumann problem, here it is allowed to have $\alpha = 0$. Recall that if $\alpha \in L^\infty(\Omega)$ with $\alpha \geq 0$, $\alpha \not\equiv 0$, the term $\alpha(x)|u|^{p-2}u$ was essential to develop the method of sub-supersolution under Neumann boundary condition (see [3]). Actually, in the Robin problem, the hypothesis $\beta(x) \geq 0$ for a.e. $x \in \partial\Omega$, $\beta \not\equiv 0$, is a substitute for the condition $\alpha(x) \geq 0$ for a.e. $x \in \Omega$, $\alpha \not\equiv 0$, assumed for the Neumann problem.

The reaction term $f(x, u, \nabla u)$ in the equation (1) is determined by a Carathéodory function $f : \Omega \times \mathbb{R} \times \mathbb{R}^N \to \mathbb{R}$, i.e., $f(\cdot, s, \xi)$ is measurable for all $(s, \xi) \in \mathbb{R} \times \mathbb{R}^N$ and $f(x, \cdot, \cdot)$ is continuous for a.e. $x \in \Omega$. This term, depending not only on the solution u but also on its gradient ∇u, is called convection. It prevents to have a variational structure for problem (1) and thus the variational methods are not applicable, which creates a serious difficulty for handling (1).

The Robin problems exhibiting convection term as is the case in (1) have only recently been studied. We refer to [4–8] for results on the existence of solutions to such problems, where the approach is based on fixed point theorems or on surjectivity criteria for monotone-type operators. We also mention that a singular Robin problem involving convection has recently been treated in [9]. There are many results for Robin problems with variational structure, thus without a convection term. In this direction, we cite, e.g., [10–15]. The aim of the present work is to study the Robin problem (1) with general gradient dependence through the method of sub-supersolution. Due to the lack of variational structure, one cannot handle such a problem by variational methods. We recall that in the study of non-variational elliptic problems one develops arguments as, for instance, the lower and upper solution method with monotone iterations, approximation approach of Galerkin-type, surjectivity theorems for monotone-type operators, fixed point theorems, topological degree theory, bifurcation theory examining phenomena as branches of solutions and blow-up. It is beyond the scope of our paper to review this huge amount of work. We only illustrate certain of these topics with a few recent references: a comparison principle and approximation process relying on a Schauder basis in [16], a fixed point approach using minimal solutions in [17], estimates based on Trudinger-Moser inequality for problems with exponential nonlinearities in [18]. We also mention the classical monographs [19,20], which are fundamental references for general elliptic equations.

According to our knowledge, this is the first time when the method of sub-supersolution is systematically implemented for nonlinear Robin problems with convection. We prove a general existence and location result for a solution to be enclosed in the ordered interval determined by a sub-supersolution. Specifically, given a subsolution $\underline{u}$ and a supersolution $\overline{u}$ for problem (1) with $\underline{u} \leq \overline{u}$ a.e. in Ω (see Section 2 for the relevant definitions), our main abstract result provides the existence of a solution u to problem (1) satisfying $\underline{u} \leq u \leq \overline{u}$ a.e. in Ω. This is an important qualitative property of the solution u offering a priori estimates. The growth condition that we suppose in the variable s for the nonlinearity $f(x, s, \xi)$ concerns only the real interval $[\underline{u}(x), \overline{u}(x)]$. We emphasize that our abstract result can be applied provided we know sub-supersolutions, i.e., ordered pairs of a subsolution $\underline{u}$ and a supersolution $\overline{u}$ for problem (1) with $\underline{u} \leq \overline{u}$ i.e., in Ω, so the task to find such ordered pairs becomes the primary task in applying the method. In this sense, we provide an application of our main result to get positive solutions for a class of nonlinear Robin problem with convection term by showing explicitly how one can effectively determine sub-supersolutions. Results, as are given here, have recently been established in [21] for nonlinear Dirichlet problems with convection and in [3] for nonlinear Neumann problems with convection. General ideas regarding the method of sub-supersolution can be found in [1,22].

The rest of the paper is organized as follows. Section 2 discusses the background needed in the sequel. Section 3 focuses on a related operator equation, which is of independent interest. Section 4 sets forth our main result. Section 5 contains our application to produce positive solutions.

2. Prerequisites of Sub-Supersolution Method

This section contains preliminaries that will be used in the sequel. First, we fix some notation. For any $r \in \mathbb{R}$, we set $r^+ = \max\{r, 0\}$ (the positive part of r). If $r > 1$, we also set $r' = \frac{r}{r-1}$ (the Hölder conjugate of r). In particular, for $p \in (1, +\infty)$ we have $p' = \frac{p}{p-1}$.

As indicated in Section 1, Ω is a bounded domain in $\mathbb{R}^N$ with $N \geq 3$ whose boundary $\partial\Omega$ is of class C^1. In order to avoid repetitive arguments, we suppose that $N > p$. The complementary case $N \leq p$ can be treated along the same lines and actually is easier. By $\|\cdot\|_{L^r(\Omega)}$ we denote the usual norm on the Banach space $L^r(\Omega)$.

We seek the solutions to problem (1) in the Sobolebv space $W^{1,p}(\Omega)$, which is a Banach space equipped with the norm

$$\|u\|_{1,p} := \left(\|u\|^p_{L^p(\Omega)} + \|\nabla u\|^p_{L^p(\Omega)}\right)^{\frac{1}{p}}.$$

For our study of problem (1) it is convenient to use the following equivalent norm on $W^{1,p}(\Omega)$ (see, e.g., ([23], Lemma 2.7) or ([15], Proposition 2.8))

$$\|u\|_{\beta,1,p} := \left(\int_{\partial\Omega} \beta(\sigma)|u(\sigma)|^p d\sigma + \|\nabla u\|^p_{L^p(\Omega)}\right)^{\frac{1}{p}}. \tag{2}$$

The dual space of $W^{1,p}(\Omega)$ is denoted $(W^{1,p}(\Omega))^*$, while the notation $\langle\cdot,\cdot\rangle$ designates the duality pairing between $W^{1,p}(\Omega)$ and $(W^{1,p}(\Omega))^*$, we denote by $\to$ the strong convergence and by $\rightharpoonup$ the weak convergence. The Sobolev embedding theorem ensures that the space $W^{1,p}(\Omega)$ is continuously embedded in $L^{p^*}(\Omega)$, where p^* is the Sobolev critical exponent $p^* = \frac{Np}{N-p}$ (we have supposed $N > p$). Moreover, by the Rellich–Kondrachov theorem, $W^{1,p}(\Omega)$ is compactly embedded in $L^r(\Omega)$ for every $r \in [1, p^*)$.

Corresponding to the map $A : \overline{\Omega} \times \mathbb{R}^N \to \mathbb{R}^N$ describing the principal part of the equation in problem (1), we introduce the operator $\tilde{A} : W^{1,p}(\Omega) \to (W^{1,p}(\Omega))^*$ defined by

$$\langle \tilde{A}(u), v\rangle = \int_\Omega A(x, \nabla u) \cdot \nabla v dx \text{ for all } u, \tag{3}$$

which is well defined thanks to assumption $(A1)$. It turns out from assumption $(A2)$ and the continuity of A that $A(x, \xi)$ is maximal monotone in the variable ξ for all $x \in \Omega$. This allows us to invoke ([2], Proposition 10), which yields:

Proposition 1. *Assume that the continuous map $A : \overline{\Omega} \times \mathbb{R}^N \to \mathbb{R}^N$ satisfies the conditions $(A1)$ and $(A2)$. Then the map $\tilde{A} : W^{1,p}(\Omega) \to (W^{1,p}(\Omega))^*$ in (3) has the (S_+)-property, that is, any sequence $\{u_n\} \subset W^{1,p}(\Omega)$ with $u_n \rightharpoonup u$ in $W^{1,p}(\Omega)$ and $\limsup_{n\to+\infty}\langle \tilde{A}(u_n), u_n - u\rangle \leq 0$ fulfills $u_n \to u$ in $W^{1,p}(\Omega)$.*

There exists a unique continuous linear map $\gamma : W^{1,p}(\Omega) \to L^p(\partial\Omega)$ called the trace map such that

$$\gamma(u) = u_{|\partial\Omega} \text{ for all } u \in W^{1,p}(\Omega) \cap C(\overline{\Omega}).$$

The kernel of $\gamma : W^{1,p}(\Omega) \to L^p(\partial\Omega)$ is $W^{1,p}_0(\Omega)$. Recalling that $N > p$, the trace map γ is compact from $W^{1,p}(\Omega)$ into $L^\eta(\partial\Omega)$ for all $\eta \in [1, \frac{(N-1)p}{N-p})$ (see, e.g., ([22], Theorem 2.79)). As usual, we drop the notation of the trace map γ writing simply u in place of $\gamma(u)$. The co-normal derivative $A(x, \nabla u) \cdot \nu(x)$, appearing in the boundary condition in problem (1), is obtained by extending the map $u(\cdot) \mapsto A(\cdot, \nabla u(\cdot)) \cdot \nu(\cdot)$, from $C^1(\overline{\Omega})$ to $W^{1,p}(\Omega)$.

By a (weak) solution to problem (1) we mean a function $u \in W^{1,p}(\Omega)$ such that $f(x,u,\nabla u) \in L^{(p^*)'}(\Omega)$ and

$$\int_\Omega A(x,\nabla u)\cdot\nabla v\,dx + \int_\Omega \alpha(x)|u|^{p-2}uv\,dx + \int_{\partial\Omega}\beta(x)|u|^{p-2}uv d\sigma = \int_\Omega f(x,u,\nabla u)v\,dx \tag{4}$$

for all $v \in W^{1,p}(\Omega)$.

A function $\underline{u} \in W^{1,p}(\Omega)$ is called a subsolution for problem (1) if $f(\cdot,\overline{u}(\cdot),\nabla\overline{u}(\cdot)) \in L^{(p^*)'}(\Omega)$ and

$$\int_\Omega (A(x,\nabla\underline{u})\cdot\nabla v + \alpha(x)|\underline{u}|^{p-2}\underline{u}v)dx + \int_{\partial\Omega}\beta(x)|\underline{u}|^{p-2}\underline{u}v d\sigma \le \int_\Omega f(x,\underline{u},\nabla\underline{u})v dx, \tag{5}$$

for all $v \in W^{1,p}(\Omega)$, $v \ge 0$ a.e. in Ω.

Symmetrically, a function $\overline{u} \in W^{1,p}(\Omega)$ is called a supersolution for problem (1) if $f(\cdot,\overline{u}(\cdot),\nabla\overline{u}(\cdot)) \in L^{(p^*)'}(\Omega)$ and

$$\int_\Omega (A(x,\nabla\overline{u})\cdot\nabla v + \alpha(x)|\overline{u}|^{p-2}\overline{u}v)\,dx + \int_{\partial\Omega}\beta(x)|\overline{u}|^{p-2}\overline{u}v d\sigma \ge \int_\Omega f(x,\overline{u},\nabla\overline{u})v\,dx, \tag{6}$$

for all $v \in W^{1,p}(\Omega)$, $v \ge 0$ a.e. in Ω.

Due to assumption $(A1)$, the integrals in the above definitions exist. We notice that $u \in W^{1,p}(\Omega)$ is a solution of (1) if and only if u is simultaneously a subsolution and a supersolution.

We are going to argue with a sub-supersolution for problem (1), that is, an ordered pair of a subsolution $\underline{u}$ and a supersolution $\overline{u}$ such that $\underline{u} \le \overline{u}$, which means the pointwise inequality $\underline{u}(x) \le \overline{u}(x)$ for a.e. $x \in \Omega$. Then we can associate the ordered interval

$$[\underline{u},\overline{u}] = \{w \in W^{1,p}(\Omega) : \underline{u} \le w \le \overline{u}\}.$$

Our goal is to obtain a solution $u \in W^{1,p}(\Omega)$ of problem (1) with the location property $u \in [\underline{u},\overline{u}]$, which will be achieved through comparison by means of a truncation operator that we now describe. Corresponding to a subsolution $\underline{u}$ and a supersolution $\overline{u}$ satisfying $\underline{u} \le \overline{u}$ a.e. in Ω, we define the truncation operator $T = T(\underline{u},\overline{u}) : W^{1,p}(\Omega) \to W^{1,p}(\Omega)$ by

$$T(u)(x) = \begin{cases} \underline{u}(x) & \text{if} \quad u(x) < \underline{u}(x) \\ u(x) & \text{if} \quad \underline{u}(x) \le u(x) \le \overline{u}(x) \\ \overline{u}(x) & \text{if} \quad u(x) > \overline{u}(x) \end{cases} \tag{7}$$

for all $u \in W^{1,p}(\Omega)$ and a.e. $x \in \Omega$. It readily follows that $T : W^{1,p}(\Omega) \to W^{1,p}(\Omega)$ is continuous and bounded (in the sense that it maps bounded sets into bounded sets).

We shall also need the (negative) Dirichlet p-Laplacian, which is the operator $-\Delta_p : W_0^{1,p}(\Omega) \to W^{-1,p'}(\Omega) = (W_0^{1,p}(\Omega))^*$ given by

$$\langle -\Delta_p u, v\rangle = \int_\Omega |\nabla u|^{p-2}\nabla u\cdot\nabla v dx \text{ for all } u,v \in W_0^{1,p}(\Omega).$$

It is well-known (see, e.g., ([1], Proposition 9.47)) that there exists a least positive number $\lambda_1 > 0$ (called the first eigenvalue of $-\Delta_p$) for which the Dirichlet problem

$$\begin{cases} -\Delta_p\varphi_1 = \lambda_1|\varphi_1|^{p-2}\varphi_1 & \text{in } \Omega \\ \varphi_1 = 0 & \text{on } \partial\Omega \end{cases} \tag{8}$$

has a nontrivial solution $\varphi_1 \in W_0^{1,p}(\Omega)$. By the regularity theory we have $\varphi_1 \in C^1(\overline{\Omega})$. Moreover, we can choose φ_1 to satisfy $\varphi_1 > 0$ in Ω.

Finally, we mention a few things about the pseudomonotone operators. Let X be a Banach space with the norm $\|\cdot\|$ and its dual X^*. We denote by $\langle\cdot,\cdot\rangle$ the duality pairing between X and X^*. A map $\mathcal{A}: X \to X^*$ is called bounded if it maps bounded sets into bounded sets. The map $\mathcal{A}: X \to X^*$ is said to be coercive if

$$\lim_{\|u\|\to+\infty} \frac{\langle \mathcal{A}(u), u\rangle}{\|u\|} = +\infty.$$

The map $\mathcal{A}: X \to X^*$ is called pseudomonotone if for each sequence $(u_n) \subset X$ satisfying $u_n \rightharpoonup u$ in X and $\limsup_{n\to\infty} \langle \mathcal{A}(u_n), u_n - u\rangle \le 0$, it holds

$$\langle \mathcal{A}(v), u - v\rangle \le \liminf_{n\to\infty} \langle \mathcal{A}(u_n), u_n - v\rangle \quad \text{for all } v \in X.$$

The main theorem for pseudomonotone operators reads as follows (see, e.g., ([22], Theorem 2.99)).

Theorem 1. *Let X be a reflexive Banach space. If $\mathcal{A}: X \to X^*$ is a pseudomonotone, bounded and coercive map, then $\mathcal{A}$ is surjective.*

3. The Associated Operator Equation

Assume that a subsolution $\underline{u}$ and a supersolution $\overline{u}$ for problem (1) with $\underline{u} \le \overline{u}$ are given and that $f: \Omega \times \mathbb{R} \times \mathbb{R}^N \to \mathbb{R}$ satisfies the following growth condition adapted to the ordered interval $[\underline{u}, \overline{u}]$:

(H) There exist a function $\sigma \in L^{r'}(\Omega)$ with $r \in (1, p^*)$ and constants $a > 0$ and $r_1 \in (0, \frac{p}{(p^*)'})$ such that

$$|f(x,s,\xi)| \le \sigma(x) + a|\xi|^{r_1} \quad \text{for a.e. } x \in \Omega, \text{ all } s \in [\underline{u}(x), \overline{u}(x)],\ \xi \in \mathbb{R}^N.$$

We introduce the cut-off function $\pi: \Omega \times \mathbb{R} \to \mathbb{R}$ defined by

$$\pi(x,s) = \begin{cases} -(\underline{u}(x) - s)^{\frac{r_1}{p-r_1}} & \text{if } s < \underline{u}(x), \\ 0 & \text{if } \underline{u}(x) \le s \le \overline{u}(x), \\ (s - \overline{u}(x))^{\frac{r_1}{p-r_1}} & \text{if } s > \overline{u}(x), \end{cases} \tag{9}$$

where $r_1 > 0$ is the constant postulated in hypothesis (H). From (9) and the fact that $\underline{u}, \overline{u} \in L^{p^*}(\Omega)$ we infer that π verifies the growth condition

$$|\pi(x,s)| \le c|s|^{\frac{r_1}{p-r_1}} + \varrho(x) \text{ for a.e. } x \in \Omega, \text{ all } s \in \mathbb{R}, \tag{10}$$

with a constant $c > 0$ and a function $\varrho \in L^{\frac{p^*(p-r_1)}{r_1}}(\Omega)$.

Now for every $\lambda > 0$ we define the nonlinear operator $A_\lambda: W^{1,p}(\Omega) \to (W^{1,p}(\Omega))^*$ by

$$\begin{aligned} \langle A_\lambda(u), v\rangle &= \int_\Omega A(x, \nabla u)\cdot\nabla v\,dx + \int_\Omega \alpha(x)|u|^{p-2}uv\,dx + \int_{\partial\Omega} \beta(x)|u|^{p-2}uv\,d\sigma \\ &+ \lambda \int_\Omega \pi(x,u)v\,dx - \int_\Omega f(x, Tu, \nabla Tu)v\,dx \quad \text{for all } u,\ v \in W^{1,p}(\Omega). \end{aligned} \tag{11}$$

Hypothesis (H) guarantees that the operator A_λ in (11) is well defined.

Due to (10), we may consider the Nemytskij operator $\Pi: L^{p^*}(\Omega) \to L^{\frac{p^*(p-r_1)}{r_1}}(\Omega)$, associated to the function π in (10), namely $\Pi(u) = \pi(\cdot, u(\cdot))$ for all $u \in L^{p^*}(\Omega)$. It is well defined, continuous and bounded. The condition in (H) that $r_1 < \frac{p}{(p^*)'}$ is equivalent to $\frac{p^*(p-r_1)}{r_1} > (p^*)'$. Hence, by the Rellich–Kondrachov compact embedding theorem, the Nemytskij operator $\Pi: W^{1,p}(\Omega) \to (W^{1,p}(\Omega))^*$ is completely continuous.

Thanks to hypothesis (H) we also have the Nemytskij operator $N_f : [\underline{u}, \overline{u}] \to (W^{1,p}(\Omega))^*$ on the ordered interval $[\underline{u}, \overline{u}]$ which is associated to the function $f : \Omega \times \mathbb{R} \times \mathbb{R}^N \to \mathbb{R}$, that is

$$\langle N_f(u), v\rangle = \int_\Omega f(x, u(x), \nabla u(x))v(x)\, dx$$

for all $u \in [\underline{u}, \overline{u}]$ and $v \in W^{1,p}(\Omega)$. Using (H) we see that $f(\cdot, u(\cdot), \nabla u(\cdot)) \in L^{\frac{p}{r_1}}(\Omega)$. As $v \in L^{p^*}(\Omega)$ and $\frac{p}{r_1} > (p^*)'$, the above integral exists. By virtue of the strict inequality $\frac{p}{r_1} > (p^*)'$, the Rellich–Kondrachov compact embedding theorem implies that the Nemytskij operator $N_f : [\underline{u}, \overline{u}] \to (W^{1,p}(\Omega))^*$ is completely continuous.

Again through the Rellich–Kondrachov compact embedding theorem we can show that the operator $B : W^{1,p}(\Omega) \to (W^{1,p}(\Omega))^*$ given by

$$\langle B(u), v\rangle = \int_\Omega \alpha(x)|u(x)|^{p-2}u(x)v(x)\, dx$$

for all $u, v \in W^{1,p}(\Omega)$ is completely continuous.

Consider also the operator $\Gamma : W^{1,p}(\Omega) \to (W^{1,p}(\Omega))^*$ given by

$$\langle \Gamma(u), v\rangle = \int_{\partial\Omega} \beta(\sigma)|u(\sigma)|^{p-2}u(\sigma)v(\sigma)d\sigma \tag{12}$$

for all $u, v \in W^{1,p}(\Omega)$, where the integration is done with respect to the $(N-1)$-dimensional Hausdorff (surface) measure on $\partial\Omega$.

Let us check that the map $\Gamma : W^{1,p}(\Omega) \to (W^{1,p}(\Omega))^*$ is completely continuous. To this end, let $u_n \rightharpoonup u$ in $W^{1,p}(\Omega)$. Then the compactness of the trace map $\gamma : W^{1,p}(\Omega) \to L^p(\partial\Omega)$ ensures the strong convergence $u_n \equiv \gamma(u_n) \to u \equiv \gamma(u)$ in $L^p(\partial\Omega)$, thus the strong convergence $|u_n|^{p-2}u_n \to |u|^{p-2}u$ in $L^{p'}(\partial\Omega)$. Taking into account (12) we deduce that $\Gamma(u_n) \to \Gamma(u)$ in $(W^{1,p}(\Omega))^*$, so $\Gamma : W^{1,p}(\Omega) \to (W^{1,p}(\Omega))^*$ is completely continuous.

For every $\lambda > 0$, the operator $A_\lambda : W^{1,p}(\Omega) \to (W^{1,p}(\Omega))^*$ in (11) has the expression

$$A_\lambda = \tilde{A} + B + \Gamma + \lambda\Pi - N_f \circ T. \tag{13}$$

The composition $N_f \circ T$ makes sense because T takes values in the ordered interval $[\underline{u}, \overline{u}]$ as seen from (7). The following theorem asserts the solvability of the equation

$$A_\lambda(u) = 0. \tag{14}$$

Theorem 2. *Assume that the conditions $(A1)$, $(A2)$ and (H) are satisfied. Then Equation (14) possesses at least a solution $u \in W^{1,p}(\Omega)$ provided $\lambda > 0$ is sufficiently large.*

Proof. In order to prove the solvability of operator Equation (14) we apply Theorem 1. We have to prove that the operator $A_\lambda : W^{1,p}(\Omega) \to (W^{1,p}(\Omega))^*$ in (13) is bounded, pseudomonotone and coercive.

By (3) and hypothesis $(A1)$, in conjunction with Hölder's inequality and the Sobolev embedding theorem, we find that

$$
\begin{aligned}
\|\tilde{A}(u)\|^{p}_{(W^{1,p}(\Omega))^*} &= \sup_{\|v\|_{\beta,1,p}\leq 1} |\langle \tilde{A}(u), v\rangle| \\
&= \sup_{\|v\|_{\beta,1,p}\leq 1} \left| \int_\Omega A(x, \nabla u)\cdot \nabla v dx \right| \\
&\leq c_2 \sup_{\|v\|_{\beta,1,p}\leq 1} \int_\Omega (|\nabla u|^{p-1} + 1)|\nabla v| dx \\
&\leq C(\|u\|^{p-1}_{\beta,1,p} + 1)
\end{aligned}
$$

for all $u \in W^{1,p}(\Omega)$, with a constant $C > 0$. This shows that the operator $\tilde{A} : W^{1,p}(\Omega) \to (W^{1,p}(\Omega))^*$ is bounded.

The composed operator $N_f \circ T$ is bounded because T is bounded and N_f is completely continuous. Since B, Π and Γ are completely continuous, it follows from (13) that $A_\lambda : W^{1,p}(\Omega) \to (W^{1,p}(\Omega))^*$ is bounded.

We claim that $A_\lambda : W^{1,p}(\Omega) \to (W^{1,p}(\Omega))^*$ is a pseudomonotone operator. Let a sequence $\{u_n\} \subset W^{1,p}(\Omega)$ satisfy $u_n \rightharpoonup u$ in $W^{1,p}(\Omega)$ and

$$
\limsup_{n\to\infty} \langle A_\lambda(u_n), u_n - u\rangle \leq 0. \tag{15}
$$

The complete continuity of the operators B, Π and Γ yields the strong convergent sequences $B(u_n) \to B(u)$, $\Pi(u_n) \to \Pi(u)$ and $\Gamma(u_n) \to \Gamma(u)$ in $(W^{1,p}(\Omega))^*$. This results in

$$
\begin{aligned}
&\lim_{n\to\infty} \langle B(u_n), u_n - v\rangle = \langle B(u), u - v\rangle, \ \lim_{n\to\infty} \langle \Pi(u_n), u_n - v\rangle = \langle \Pi(u), u - v\rangle, \\
&\lim_{n\to\infty} \langle \Gamma(u_n), u_n - u\rangle = \langle \Gamma(u), u - v\rangle
\end{aligned} \tag{16}
$$

for all $v \in W^{1,p}(\Omega)$. We infer that

$$
\lim_{n\to\infty} \langle B(u_n), u_n - u\rangle = \lim_{n\to\infty} \langle \Pi(u_n), u_n - u\rangle = \lim_{n\to\infty} \langle \Gamma(u_n), u_n - u\rangle = 0,
$$

so (15) reduces to

$$
\limsup_{n\to\infty} \langle \tilde{A}(u_n), u_n - u\rangle \leq 0. \tag{17}
$$

Inequality (17) enables us to apply Proposition 1 ensuring that the strong convergence $u_n \to u$ in $W^{1,p}(\Omega)$ holds.

At this point, we know that the strong convergence $\nabla(u_n) \to \nabla(u)$ holds in $(L^p(\Omega))^N$, so the second inequality in $(A1)$ entails $A(\cdot, \nabla u_n(\cdot)) \to A(\cdot, \nabla u(\cdot))$ strongly in $(L^{p'}(\Omega))^N$. Then for each $v \in W^{1,p}(\Omega)$ one has

$$
\begin{aligned}
\lim_{n\to\infty} \langle \tilde{A}(u_n), u_n - v\rangle &= \lim_{n\to\infty} \int_\Omega A(x, \nabla u_n)\cdot \nabla(u_n - v) dx \\
&= \int_\Omega A(x, \nabla u)\cdot \nabla(u - v) dx \\
&= \langle \tilde{A}(u), u - v\rangle.
\end{aligned} \tag{18}
$$

Taking into account of (13), (16) and (18), we arrive at

$$
\lim_{n\to\infty} \langle A_\lambda(u_n), u_n - v\rangle = \langle A_\lambda(u), u - v\rangle
$$

for all $v \in W^{1,p}(\Omega)$ and $\lambda > 0$. Therefore the operator $A_\lambda : W^{1,p}(\Omega) \to (W^{1,p}(\Omega))^*$ is pseudomonotone.

Next we show that the operator $A_\lambda : W^{1,p}(\Omega) \to (W^{1,p}(\Omega))^*$ is coercive whenever $\lambda > 0$ is sufficiently large.

Since $\alpha \in L^\infty(\Omega)$, $\alpha \geq 0$, from (11) we note that

$$\langle A_\lambda(u), u\rangle \geq \langle \tilde{A}(u), u\rangle + \int_{\partial\Omega} \beta(\sigma)|u(\sigma)|^p \, d\sigma + \lambda \int_\Omega \pi(x,u)u \, dx - \int_\Omega f(x, Tu, \nabla(Tu))u \, dx \tag{19}$$

for all $u \in W^{1,p}(\Omega)$. We estimate from below the terms in the right-hand side of (19). Assumption $(A1)$ and (3) yield

$$\langle \tilde{A}u, u\rangle \geq c_1 \|\nabla u\|^p_{L^p(\Omega)} \text{ for all } u \in W^{1,p}(\Omega). \tag{20}$$

From (9) we derive that

$$\int_\Omega \pi(x, u(x))u(x) \, dx \geq b_1 \|u\|^{\frac{p}{p-r_1}}_{L^{\frac{p}{p-r_1}}(\Omega)} - b_2 \text{ for all } u \in W^{1,p}(\Omega), \tag{21}$$

with positive constants b_1 and b_2 (see [3]).

In view of (7), we have that $\underline{u} \leq Tu \leq \overline{u}$ a.e. in Ω whenever $u \in W^{1,p}(\Omega)$. Consequently, we may set $s = (Tu)(x)$ in the statement of hypothesis (H). Then, for each $\varepsilon > 0$, we obtain through Hölder's and Young's inequalities and the Sobolev embedding theorem the estimate

$$\begin{aligned} \left| \int_\Omega f(x, Tu, \nabla(Tu))u \, dx \right| &\leq \int_\Omega (\sigma|u| + a|\nabla(Tu)|^{r_1}|u|) \, dx \\ &\leq \varepsilon \|\nabla u\|^p_{L^p(\Omega)} + c(\varepsilon)\|u\|^{\frac{p}{p-r_1}}_{L^{\frac{p}{p-r_1}}(\Omega)} + d\|u\|_{\beta,1,p}, \end{aligned} \tag{22}$$

with positive constants $c(\varepsilon)$ (depending on ε) and d.

Gathering (19)–(22) leads to

$$\langle A_\lambda(u), u\rangle \geq (c_1 - \varepsilon)\|\nabla u\|^p_{L^p(\Omega)} + \textstyle\int_{\partial\Omega} \beta(\sigma)|u(\sigma)|^p \, d\sigma + (\lambda b_1 - c(\varepsilon))\|u\|^{\frac{p}{p-r_1}}_{L^{\frac{p}{p-r_1}}(\Omega)} - d\|u\|_{\beta,1,p} - \lambda b_2 \tag{23}$$

for all $u \in W^{1,p}(\Omega)$ and $\lambda > 0$. Now we fix ε and λ to verify $\varepsilon \in (0, c_1)$ and $\lambda > \frac{c(\varepsilon)}{b_1}$. From (2) and (23) it is clear that

$$\langle A_\lambda(u), u\rangle \geq c_0\|u\|^p_{\beta,1,p} - d\|u\|_{\beta,1,p} - \lambda b_2$$

for all $u \in W^{1,p}(\Omega)$, with a constant $c_0 > 0$. Due to the fact that $p > 1$, it turns out

$$\lim_{\|u\|_{\beta,1,p} \to +\infty} \frac{\langle A_\lambda(u), u\rangle}{\|u\|_{\beta,1,p}} = +\infty,$$

thereby the operator A_λ is coercive.

Summarizing, we have proved that the operator $A_\lambda : W^{1,p}(\Omega) \to (W^{1,p}(\Omega))^*$ is bounded, pseudomonotone and coercive. This allows us to apply Theorem 1 with $\mathcal{A} = A_\lambda$ for $\lambda > 0$ sufficiently large. The surjectivity of A_λ implies the existence of a solution $u \in W^{1,p}(\Omega)$ of Equation (14), thus completing the proof. □

Remark 1. *As a consequence of* (23)*, we can precisely determine the threshold of* $\lambda > 0$ *in the statement of Theorem* 2*.*

4. Main Abstract Result for Problem (1)

Our result regarding the method of sub-supersolution for problem (1) is stated as follows.

Theorem 3. *Assume that the conditions* $(A1)$, $(A2)$ *and* (H) *are satisfied. Then problem* (P) *possesses a solution* $u \in W^{1,p}(\Omega)$ *satisfying* $\underline{u} \leq u \leq \overline{u}$ *a.e. in* Ω, *where* $\underline{u}$ *and* $\overline{u}$ *are the subsolution and the supersolution that are postulated in assumption* (H).

Proof. According to Theorem 2 we can fix $\lambda > 0$ sufficiently large such that equation (14) admits a solution $u \in W^{1,p}(\Omega)$. Explicitly, this reads as

$$
\begin{aligned}
&\langle \tilde{A}(u), v\rangle + \int_\Omega \alpha(x)|u|^{p-2}uv dx + \lambda \int_\Omega \pi(x,u)v dx + \int_{\partial\Omega} \beta(x)|u|^{p-2}uv d\sigma \\
&= \int_\Omega f(x, Tu, \nabla(Tu))v dx \text{ for all } v \in W^{1,p}(\Omega).
\end{aligned}
\tag{24}
$$

Let us prove that $u \leq \overline{u}$ a.e. in Ω. Inserting $v = (u - \overline{u})^+ \in W^{1,p}(\Omega)$ in (6) and (24) renders

$$
\begin{aligned}
&\langle \tilde{A}(\overline{u}), (u-\overline{u})^+\rangle + \int_\Omega \alpha(x)|\overline{u}|^{p-2}\overline{u}(u-\overline{u})^+ dx + \int_{\partial\Omega} \beta(x)|\overline{u}|^{p-2}\overline{u}(u-\overline{u})^+ d\sigma \\
&\geq \int_\Omega f(x, \overline{u}, \nabla\overline{u})(u-\overline{u})^+ dx
\end{aligned}
\tag{25}
$$

and

$$
\begin{aligned}
&\langle \tilde{A}(u), (u-\overline{u})^+\rangle + \int_\Omega \alpha(x)|u|^{p-2}u(u-\overline{u})^+ dx + \lambda \int_\Omega \pi(x,u)(u-\overline{u})^+ dx \\
&+ \int_{\partial\Omega} \beta(x)|u|^{p-2}u(u-\overline{u})^+ d\sigma \\
&= \int_\Omega f(x, Tu, \nabla(Tu))(u-\overline{u})^+ dx.
\end{aligned}
\tag{26}
$$

Subtract (25) from (26) and use (3) and (7) to deduce that

$$
\begin{aligned}
&\int_\Omega (A(x,\nabla u) - A(x,\nabla\overline{u}))\nabla(u-\overline{u})^+ dx + \int_{\partial\Omega} \beta(x)(|u|^{p-2}u - |\overline{u}|^{p-2}\overline{u})(u-\overline{u})^+ d\sigma \\
&+ \int_\Omega \alpha(x)(|u|^{p-2}u - |\overline{u}|^{p-2}\overline{u})(u-\overline{u})^+ dx + \lambda \int_\Omega \pi(x,u)(u-\overline{u})^+ dx \\
&\leq \int_\Omega (f(x, Tu, \nabla(Tu)) - f(x, \overline{u}, \nabla\overline{u}))(u-\overline{u})^+ \, dx \\
&= \int_{\{u>\overline{u}\}} (f(x, Tu, \nabla(Tu)) - f(x, \overline{u}, \nabla\overline{u}))(u-\overline{u}) dx = 0.
\end{aligned}
\tag{27}
$$

The monotonicity of $A(x, \cdot)$, guaranteed by assumption (A_2), and the monotonicity of the map $\xi \mapsto |\xi|^{p-2}\xi$ on $\mathbb{R}^N$ give

$$
\begin{aligned}
&\int_\Omega (A(x,\nabla u) - A(x,\nabla\overline{u}))\nabla(u-\overline{u})^+ dx \\
&= \int_{\{u>\overline{u}\}} (A(x,\nabla u) - A(x,\nabla\overline{u}))(\nabla u - \nabla\overline{u}) dx \geq 0,
\end{aligned}
$$

$$
\begin{aligned}
&\int_\Omega \alpha(x)(|u|^{p-2}u - |\overline{u}|^{p-2}\overline{u})(u-\overline{u})^+ dx \\
&= \int_{\{u>\overline{u}\}} \alpha(x)(|u|^{p-2}u - |\overline{u}|^{p-2}\overline{u})(u-\overline{u}) dx \geq 0,
\end{aligned}
$$

$$
\begin{aligned}
&\int_{\partial\Omega} \beta(\sigma)(|u|^{p-2}u - |\overline{u}|^{p-2}\overline{u})(u-\overline{u})^+ d\sigma \\
&= \int_{\{\sigma\in\partial\Omega \,:\, u>\overline{u}\}} \beta(\sigma)(|u|^{p-2}u - |\overline{u}|^{p-2}\overline{u})(u-\overline{u}) d\sigma \geq 0.
\end{aligned}
$$

From (27) and (9) we obtain

$$\int_{\{u>\overline{u}\}}(u-\overline{u})^{\frac{p}{p-r_1}}dx=\int_{\Omega}\pi(x,u)(u-\overline{u})^{+}dx\leq 0,$$

where $u \leq \overline{u}$ a.e in Ω.

Next we show that $\underline{u} \leq u$ a.e in Ω. Setting $v=(\underline{u}-u)^{+}\in W^{1,p}(\Omega)$ in (5) and (24) produces

$$\begin{aligned}&\langle \tilde{A}(\underline{u}),(\underline{u}-u)^{+}\rangle+\int_{\Omega}\alpha(x)|\underline{u}|^{p-2}\underline{u}(\underline{u}-u)^{+}dx+\int_{\partial\Omega}\beta(x)|\underline{u}|^{p-2}\underline{u}(\underline{u}-u)^{+}d\sigma\\&\leq\int_{\Omega}f(x,\underline{u},\nabla\underline{u})(\underline{u}-u)^{+}dx\end{aligned}\tag{28}$$

and

$$\begin{aligned}&\langle \tilde{A}(u),(\underline{u}-u)^{+}\rangle+\int_{\Omega}\alpha(x)|u|^{p-2}u(\underline{u}-u)^{+}dx+\lambda\int_{\Omega}\pi(x,u)(\underline{u}-u)^{+}dx\\&+\int_{\partial\Omega}\beta(x)|u|^{p-2}u(\underline{u}-u)^{+}d\sigma\\&=\int_{\Omega}f(x,Tu,\nabla(Tu))(\underline{u}-u)^{+}dx.\end{aligned}\tag{29}$$

By subtracting (29) from (28) and taking into account (3) we arrive at

$$\begin{aligned}&\int_{\Omega}(A(x,\nabla\underline{u})-A(x,\nabla u))\nabla(\underline{u}-u)^{+}dx+\int_{\partial\Omega}\beta(x)(|\underline{u}|^{p-2}\underline{u}-|u|^{p-2}u)(\underline{u}-u)^{+}d\sigma\\&+\int_{\Omega}\alpha(x)(|\underline{u}|^{p-2}\underline{u}-|u|^{p-2}u)(\underline{u}-u)^{+}dx-\lambda\int_{\Omega}\pi(x,u)(\underline{u}-u)^{+}dx\\&\leq\int_{\Omega}(f(x,\underline{u},\nabla\underline{u})-f(x,Tu,\nabla(Tu))(\underline{u}-u)^{+}dx\\&=\int_{\{\underline{u}>u\}}(f(x,\underline{u},\nabla\underline{u})-f(x,Tu,\nabla(Tu)))(\underline{u}-u)^{+}dx=0.\end{aligned}\tag{30}$$

Along (9) and proceeding as above, (30) results in

$$-\int_{\{\underline{u}>u\}}-(\underline{u}-u)^{\frac{p}{p-r_1}}dx=-\int_{\Omega}\pi(x,u)(\underline{u}-u)^{+}dx\leq 0,$$

which entails that $\underline{u} \leq u$ a.e in Ω, thus proving the claim.

Therefore the solution $u \in W^{1,p}(\Omega)$ of the operator equation (14) verifies the enclosure property $\underline{u} \leq u \leq \overline{u}$ a.e. in Ω. Then we obtain from (7) and (9) that $Tu = u$ and $\Pi(u) = 0$. Hence for our function u the equalities (24) and (4) coincide. We see that $u \in W^{1,p}(\Omega)$ is a solution of the original problem (1) fulfilling in addition $\underline{u} \leq u \leq \overline{u}$ a.e. in Ω. This completes the proof. □

5. An Application

The aim of this section is to apply Theorem 3 to establish the existence of positive solutions of Robin problem (1). The main point is to find appropriate ordered sub-supersolutions. The approach can be used to get other types of solutions.

In order to simplify the presentation, we focus on problem (1) driven by the Robin p-Laplacian, $1 < p < +\infty$, and when $\alpha(x) \equiv 0$ and the x-dependence in the convection term $f(x,s,\xi)$ is dropped. We emphasize that $\alpha \equiv 0$ marks a sharp distinction in regard to the Neumann problem. Specifically, we consider the (purely) Robin problem

$$\begin{cases}-\Delta_p u=f(u,\nabla u) & \text{in } \Omega\\ |\nabla u|^{p-2}\nabla u\cdot\nu(x)+\beta(x)|u|^{p-2}u=0 & \text{on } \partial\Omega,\end{cases}\tag{31}$$

with $\beta(x) \geq 0$ for a.e. $x \in \partial\Omega$, $\beta \not\equiv 0$.

We suppose that $f : \mathbb{R} \times \mathbb{R}^N \to \mathbb{R}$ is a continuous function verifying the following assumption:

(H') There exist constants $a_0 > 0, a_1 > 0, b > 0$ and $r_1 \in (0, \frac{p}{(p^*)'})$ such that

$$|f(s,\xi)| \leq a_1(1 + |\xi|^{r_1}) \text{ for all } s \in (0,b],\ \xi \in \mathbb{R}^N, \tag{32}$$

$$\lambda_1 s^{p-1} \leq f(s,\xi) \text{ for all } s \in (0,a_0),\ |\xi| < a_0 \tag{33}$$

and

$$f(b,0) = 0. \tag{34}$$

The condition (33) involves the first eigenvalue λ_1 of the (negative) Dirichlet p-Laplacian as given in (8). Let us note that $u = b$ is not a solution to problem (31) because the boundary condition is not verified. We formulate the following result concerning problem (31).

Theorem 4. *Assume that the conditions* $(A1)$, $(A2)$ *and* (H') *are satisfied. Then the Robin problem* (31) *possesses a (positive) solution* $u \in W^{1,p}(\Omega)$ *satisfying* $0 < u \leq b$ *a.e. in* Ω.

Proof. Fix an eigenfunction φ_1 of $-\Delta_p$ on $W_0^{1,p}(\Omega)$, with $\varphi_1 > 0$ in Ω, corresponding to the first eigenvalue λ_1 (see (8) and the related comments). Since $\varphi_1 \in C^1(\overline{\Omega})$, we can choose an $\varepsilon > 0$ such that

$$\varepsilon\varphi_1(x) < a_0 \text{ and } \varepsilon|\nabla\varphi_1(x)| < a_0 \text{ for all } x \in \Omega, \tag{35}$$

where a_0 is the positive constant prescribed in hypothesis (H').

We note that $\underline{u} = \varepsilon\varphi_1$ is a subsolution in the sense of (5) for the Robin problem (31). Indeed, by (8), (33) and (35) and since the trace of $\underline{u}$ on $\partial\Omega$ vanishes, we infer that

$$\begin{aligned}
\int_\Omega |\nabla\underline{u}|^{p-2}\nabla\underline{u}\cdot\nabla v dx + \int_{\partial\Omega}\beta(x)|\underline{u}|^{p-2}\underline{u}v d\sigma &= \varepsilon^{p-1}\int_\Omega |\nabla\varphi_1(x)|^{p-2}\nabla\varphi_1(x)\cdot\nabla v(x)dx \\
&= \lambda_1\int_\Omega (\varepsilon\varphi_1(x))^{p-1}v(x)dx \\
&\leq \int_\Omega f(\varepsilon\varphi_1(x), \varepsilon\nabla\varphi_1(x))v(x)dx \\
&= \int_\Omega f(\underline{u}(x), \nabla\underline{u}(x))v(x)dx \text{ for all } v \in W^{1,p}(\Omega),\ v \geq 0.
\end{aligned}$$

This proves that $\underline{u} = \varepsilon\varphi_1$ is a subsolution of problem (31).

Now we observe that the constant function $\overline{u} = b$ is a supersolution of problem (31). Indeed, let us notice from assumption (34) that

$$\begin{aligned}
\int_\Omega |\nabla\overline{u}|^{p-2}\nabla\overline{u}\cdot\nabla v dx + \int_{\partial\Omega}\beta(x)|\overline{u}|^{p-2}\overline{u}v d\sigma &= \int_{\partial\Omega}\beta(x)b^{p-1}v(x)d\sigma \\
&\geq 0 = \int_\Omega f(b,0)v(x)dx \\
&= \int_\Omega f(\overline{u}(x), \nabla\overline{u}(x))v(x)dx
\end{aligned}$$

for all $v \in W^{1,p}(\Omega)$ with $v \geq 0$, which confirms that $\overline{u} = b$ is a supersolution of problem (31) in the sense of (6).

For a possibly smaller $\varepsilon > 0$ to be fulfilled $\varepsilon\varphi_1(x) \leq b$ whenever $x \in \Omega$, the inequality $\underline{u} \leq \overline{u}$ holds true. The growth condition in (H) is satisfied due to (32) because the pointwise intervals $[\underline{u}(x), \overline{u}(x)]$ are all included in the bounded interval $(0, b]$. Altogether we are in a position to apply Theorem 3, which yields the desired conclusion. □

We provide a simple example illustrating the applicability of Theorem 4.

Example 1. *Let* $f : \mathbb{R} \times \mathbb{R}^N \to \mathbb{R}$ *be defined by*

$$f(s,\xi) = g(s) + h(\xi) \text{ for all } (s,\xi) \in \mathbb{R} \times \mathbb{R}^N,$$

with $g : \mathbb{R} \to \mathbb{R}$ *defined by*

$$g(s) = \begin{cases} 0 & \text{if } s < 0 \text{ or } s > 2 \\ \lambda_1 s^{p-1} & \text{if } 0 \le s \le 1 \\ \lambda_1 (2-s)^{p-1} & \text{if } 1 < s \le 2 \end{cases}$$

and any continuous function $h : \mathbb{R}^N \to \mathbb{R}$ *satisfying* $h(\xi) \ge 0$, $h(0) = 0$ *and*

$$0 \le h(\xi) \le a_2(1 + |\xi|^{r_1}) \text{ for all } \xi \in \mathbb{R}^N,$$

with constants $a_2 > 0$ *and* $r_1 \in (0, \frac{p}{(p^*)'})$. *We note that* $f(2,0) = 0$ *and*

$$f(s,\xi) = g(s) + h(\xi) \ge \lambda_1 s^{p-1} \text{ for all } 0 \le s \le 1,\ \xi \in \mathbb{R}^N.$$

Hypothesis (H') *is verified taking* $a_0 = 1$ *and* $b = 2$. *Theorem 4 can be applied to problem (31) with* $f(s,\xi)$ *given above.*

Author Contributions: Conceptualization, D.M., A.S. and E.T. All authors contributed equally to this paper. All authors have read and agreed to the published version of the manuscript.

Funding: This research received no external funding.

Acknowledgments: The last two authors are members of Gruppo Nazionale per l'Analisi Matematica, la Probabilità e le loro Applicazioni (GNAMPA) of Istituto Nazionale di Alta Matematica (INdAM). The paper is partially supported by PRIN 2017—Progetti di Ricerca di rilevante Interesse Nazionale, Nonlinear Differential Problems via Variational, Topological and Set-valued Methods. The authors thank the referees for careful reading and useful comments that helped to improve the paper.

Conflicts of Interest: The authors declare no conflict of interest.

References

1. Motreanu, D.; Motreanu, V.V.; Papageorgiou, N.S. *Topological and Variational Methods with Applications to Nonlinear Boundary Value Problems*; Springer: New York, NY, USA, 2014.
2. Miyajima, S.; Motreanu, D.; Tanaka, M. Multiple existence results of solutions for the Neumann problems via super- and sub-solutions. *J. Funct. Anal.* **2012**, *262*, 1921–1953. [CrossRef]
3. Motreanu, D.; Sciammetta, A.; Tornatore, E. A sub-supersolution approach for Neumann boundary value problems with gradient dependence. *Nonlinear Anal. Real World Appl.* **2020**, *54*, 103096. [CrossRef]
4. Averna, D.; Papageorgiou, N.S.; Tornatore, E. Positive solutions for nonlinear Robin problems with convection. *Math. Methods Appl. Sci.* **2019**, *42*, 1907–1920. [CrossRef]
5. Bai, Y.; Gasinski, L.; Papageorgiou, N.S. Nonlinear nonhomogeneous Robin problems with dependence on the gradient. *Bound. Value Probl.* **2018**, *17*, 1–24. [CrossRef]
6. Candito, P.; Gasinski, L.; Papageorgiou, N.S. Nonlinear nonhomogeneous Robin problems with convection. *Ann. Acad. Sci. Fenn. Math.* **2019**, *44*, 755–767. [CrossRef]
7. Faraci, F.; Motreanu, D.; Puglisi, D. Positive solutions of quasi-linear elliptic equations with dependence on the gradient. *Calc. Var. Partial Differ. Equ.* **2015**, *54*, 525–538. [CrossRef]
8. Papageorgiou, N.S.; Radulescu, V.; Repovs, D. Positive solutions for nonvariational Robin problems. *Asymptot. Anal.* **2018**, *108*, 243–255. [CrossRef]
9. Guarnotta, U.; Marano, S.A.; Motreanu, D. On a Singular Robin Problem with Convection Terms. Available online: http://arxiv.org/abs/1909.09834 (accessed on 1 March 2020).
10. Averna, D.; Papageorgiou, N.S.; Tornatore, E. Positive solutions for nonlinear Robin problems. *Electron. J. Differ. Equ.* **2017**, *204*, 1–25. [CrossRef]

11. D'Aguì, G.; Marano, S.; Papageorgiou, N.S. Multiple solutions to a Robin problem with indefinite weight and asymmetric reaction. *J. Math. Anal. Appl.* **2016**, *433*, 1821–1845. [CrossRef]
12. Guarnotta, U.; Marano, S.A.; Papageorgiou, N.S. Multiple nodal solutions to a Robin problem with sign-changing potential and locally defined reaction. *Atti Accad. Naz. Lincei Rend. Lincei Mat. Appl.* **2019**, *30*, 269–294. [CrossRef]
13. Marano, S.A.; Papageorgiou, N.S. Asymmetric Robin boundary-value problems with p-Laplacian and indefinite potential. *Electron. J. Differ. Equ.* **2018**, *2018*, 1–21.
14. Papageorgiou, N.S.; Radulescu, V.; Repovs, D. Positive solutions for perturbations of the Robin eigenvalue problem plus an indefinite potential. *Discret. Contin. Dyn. Syst.* **2017**, *37*, 2589–2618. [CrossRef]
15. Papageorgiou, N.S.; Winkert, P. Solutions with sign information for nonlinear nonhomogeneous problems. *Math. Nachr.* **2019**, *292*, 871–891. [CrossRef]
16. Faria, L.; Miyagaki, O.; Motreanu, D. Comparison and positive solutions for problems with (p,q)-Laplacian and convection term. *Proc. Edinb. Math. Soc.* **2014**, *57*, 687–698. [CrossRef]
17. Fragnelli, G.; Mugnai, D.; Papageorgiou, N.S. Robin problems for the p-Laplacian with gradient dependence. *Discret. Contin. Dyn. Syst. Ser. S* **2019**, *12*, 287–295. [CrossRef]
18. De Araujo, A.; Faria, L. Positive solutions of quasilinear elliptic equations with exponential nonlinearity combined with convection term. *J. Differ. Equ.* **2019**, *267*, 4589–4608. [CrossRef]
19. Gilbarg D.; Trudinger, N. *Elliptic Partial Differential Equations of Second Order*; Reprint of the 1998 edition, Classics in Mathematics; Springer: Berlin, Germany, 2001.
20. Ladyzhenskaya, O.; Uraltseva, N. *Linear and Quasilinear Elliptic Equations*; Academic Press: New York, NY, USA; London, UK, 1968.
21. Motreanu, D.; Tornatore, E. Location of solutions for quasilinear elliptic equations with gradient dependence. *Electron. J. Qual. Theory Differ. Equ.* **2017**, *87*, 1–10. [CrossRef]
22. Carl, S.; Le, V.K.; Motreanu, D. *Nonsmooth Variational Problems and Their Inequalities. Comparison Principles and Applications*; Springer: New York, NY, USA, 2007.
23. Colasuonno, F. ; Pucci, P.; Varga, C. Multiple solutions for an eigenvalue problem involving p-Laplacian type operator. *Nonlinear Anal.* **2012**, *75*, 4496–4512. [CrossRef]

mathematics

MDPI

Article

The Topological Transversality Theorem for Multivalued Maps with Continuous Selections

Donal O'Regan

School of Mathematics, Statistics and Applied Mathematics, National University of Ireland, Galway H91 TK33, Ireland; donal.oregan@nuigalway.ie

Received: 22 October 2019; Accepted: 14 November 2019; Published: 15 November 2019

Abstract: This paper considers a topological transversality theorem for multivalued maps with continuous, compact selections. Basically, this says, if we have two maps F and G with continuous compact selections and $F \cong G$, then one map being essential guarantees the essentiality of the other map.

Keywords: essential maps; homotopy; selections

MSC: 47H10; 54H25

1. Introduction

In this paper, we consider multivalued maps F and G with continuous, compact selections and $F \cong G$ in this setting. The topological transversality theorem will state that F is essential if and only if G is essential (essential maps were introduced by Granas [1] and extended by Precup [2], Gabor, Gorniewicz, and Slosarsk [3], and O'Regan [4,5]). For an approach to other classes of maps, we refer the reader to O'Regan [6], where one sees that $\cong$ in the appropriate class can be challenging. However, the topological transversality theorem for multivalued maps with continuous compact selections has not been considered in detail. In this paper, we present a simple result that immediately yields a topological transversality theorem in this setting. In particular, we show that, for two maps F and G with continuous compact selections and $F \cong G$, then one map being essential (or d–essential) guarantees that the other is essential (or d–essential). We also discuss these maps in the weak topology setting.

2. Topological Transversality Theorem

We will consider a class **A** of maps. Let E be a completely regular space (i.e., a Tychonoff space) and U an open subset of E.

Definition 1. *We say $f \in D(\overline{U}, E)$ if $f : \overline{U} \to E$ is a continuous, compact map; here, $\overline{U}$ denotes the closure of U in E.*

Definition 2. *We say $f \in D_{\partial U}(\overline{U}, E)$ if $f \in D(\overline{U}, E)$ and $x \neq f(x)$ for $x \in \partial U$; here, ∂U denotes the boundary of U in E.*

Definition 3. *We say $F \in A(\overline{U}, E)$ if $F : \overline{U} \to 2^E$ with $F \in \mathbf{A}(\overline{U}, E)$ and there exists a selection $f \in D(\overline{U}, E)$ of F; here, 2^E denotes the family of nonempty subsets of E.*

Remark 1. *Let Z and W be subsets of Hausdorff topological vector spaces Y_1 and Y_2 and F a multifunction. We say $F \in PK(Z, W)$ if W is convex and there exists a map $S : Z \to W$ with $Z = \cup\{int\, S^{-1}(w) : w \in W\}$, $co\,(S(x)) \subseteq F(x)$ for $x \in Z$ and $S(x) \neq \emptyset$ for each $x \in Z$; here, $S^{-1}(w) = \{z : w \in S(z)\}$. Let E be a Hausdorff topological vector space (note topological vector spaces are completely regular), U an open subset* *of E and $\overline{U}$ paracompact. In this case, we say $F \in \mathbf{A}(\overline{U}, E)$ if $F \in PK(\overline{U}, E)$ is a compact map. Now, [7] guarantees that there exists a continuous, compact selection $f : \overline{U} \to E$ of F.*

Definition 4. *We say $F \in A_{\partial U}(\overline{U}, E)$ if $F \in A(\overline{U}, E)$ and $x \notin F(x)$ for $x \in \partial U$.*

Definition 5. *We say $F \in A_{\partial U}(\overline{U}, E)$ is essential in $A_{\partial U}(\overline{U}, E)$ if for any selection $f \in D(\overline{U}, E)$ of F and any map $g \in D_{\partial U}(\overline{U}, E)$ with $f|_{\partial U} = g|_{\partial U}$ there exists a $x \in U$ with $x = g(x)$.*

Remark 2. *If $F \in A_{\partial U}(\overline{U}, E)$ is essential in $A_{\partial U}(\overline{U}, E)$ and if $f \in D(\overline{U}, E)$ is any selection of F then there exists a $x \in U$ with $x = f(x)$ (take $g = f$ in Definition 5), so in particular there exists a $x \in U$ with $x \in F(x)$.*

Definition 6. *Let $f, g \in D_{\partial U}(\overline{U}, E)$. We say $f \cong g$ in $D_{\partial U}(\overline{U}, E)$ if there exists a continuous, compact map $h : \overline{U} \times [0,1] \to E$ with $x \neq h_t(x)$ for any $x \in \partial U$ and $t \in (0,1)$ (here $h_t(x) = h(x,t)$), $h_0 = f$ and $h_1 = g$.*

Remark 3. *A standard argument guarantees that $\cong$ in $D_{\partial U}(\overline{U}, E)$ is an equivalence relation.*

Definition 7. *Let $F, G \in A_{\partial U}(\overline{U}, E)$. We say $F \cong G$ in $A_{\partial U}(\overline{U}, E)$ if for any selection $f \in D_{\partial U}(\overline{U}, E)$ (respectively, $g \in D_{\partial U}(\overline{U}, E)$) of F (respectively, of G) we have $f \cong g$ in $D_{\partial U}(\overline{U}, E)$.*

Theorem 1. *Let E be a completely regular topological space, U an open subset of E, $F \in A_{\partial U}(\overline{U}, E)$ and $G \in A_{\partial U}(\overline{U}, E)$ is essential in $A_{\partial U}(\overline{U}, E)$. In addition, suppose*

$$\begin{cases} \text{for any selection } f \in D_{\partial U}(\overline{U}, E) \text{ (respectively, } g \in D_{\partial U}(\overline{U}, E)) \\ \text{of } F \text{ (respectively, of } G) \text{ and any map } \theta \in D_{\partial U}(\overline{U}, E) \\ \text{with } \theta|_{\partial U} = f|_{\partial U} \text{ we have } g \cong \theta \text{ in } D_{\partial U}(\overline{U}, E). \end{cases} \tag{1}$$

Then, F is essential in $A_{\partial U}(\overline{U}, E)$.

Proof. Let $f \in D_{\partial U}(\overline{U}, E)$ be any selection of F and consider any map $\theta \in D_{\partial U}(\overline{U}, E)$ with $\theta|_{\partial U} = f|_{\partial U}$. We must show that there exists a $x \in U$ with $x = \theta(x)$. Let $g \in D_{\partial U}(\overline{U}, E)$ be any selection of G. Now, (1) guarantees that there exists a continuous, compact map $h : \overline{U} \times [0,1] \to E$ with $x \neq h_t(x)$ for any $x \in \partial U$ and $t \in (0,1)$ (here, $h_t(x) = h(x,t)$), $h_0 = g$ and $h_1 = \theta$. Let

$$\Omega = \left\{ x \in \overline{U} : \ x = h(x,t) \ \text{ for some } \ t \in [0,1] \right\}.$$

Now, $\Omega \neq \emptyset$ (note G is essential in $A_{\partial U}(\overline{U}, E)$) and Ω is closed (note h is continuous) and so Ω is compact (note h is a compact map). In addition, note $\Omega \cap \partial U = \emptyset$ since $x \neq h_t(x)$ for any $x \in \partial U$ and $t \in [0,1]$. Then, since E is Tychonoff, there exists a continuous map $\mu : \overline{U} \to [0,1]$ with $\mu(\partial U) = 0$ and $\mu(\Omega) = 1$. Define the map r by $r(x) = h(x, \mu(x)) = h \circ g(x)$, where $g : \overline{U} \to \overline{U} \times [0,1]$ is given by $g(x) = (x, \mu(x))$. Note that $r \in D_{\partial U}(\overline{U}, E)$ (i.e., r is a continuous compact map) with $r|_{\partial U} = g|_{\partial U}$ (note if $x \in \partial U$ then $r(x) = h(x,0) = g(x)$) so since G is essential in $A_{\partial U}(\overline{U}, E)$ there exists a $x \in U$ with $x = r(x)$ (i.e., $x = h_{\mu(x)}(x)$). Thus, $x \in \Omega$ so $\mu(x) = 1$ and thus $x = h_1(x) = \theta(x)$. □

Let E be a topological vector space. Before we prove the topological transversality theorem, we note the following:

(a) If $f, g \in D_{\partial U}(\overline{U}, E)$ with $f|_{\partial U} = g|_{\partial U}$, then $f \cong g$ in $D_{\partial U}(\overline{U}, E)$. To see this, let $h(x,t) = (1-t)\, f(x) + t\, g(x)$ and note $h : \overline{U} \times [0,1] \to E$ is a continuous, compact map with $x \neq h_t(x)$ for any $x \in \partial U$ and $t \in (0,1)$ (note $f|_{\partial U} = g|_{\partial U}$).

Theorem 2. *Let E be a topological vector space and U an open subset of E. Suppose that F and G are two maps in $A_{\partial U}(\overline{U}, E)$ with $F \cong G$ in $A_{\partial U}(\overline{U}, E)$. Now, F is essential in $A_{\partial U}(\overline{U}, E)$ if and only if G is essential in $A_{\partial U}(\overline{U}, E)$.*

Proof. Assume G is essential in $A_{\partial U}(\overline{U}, E)$. We will use Theorem 1 to show F is essential in $A_{\partial U}(\overline{U}, E)$. Let $f \in D_{\partial U}(\overline{U}, E)$ be any selection of F, $g \in D_{\partial U}(\overline{U}, E)$ be any selection of G and consider any map $\theta \in D_{\partial U}(\overline{U}, E)$ with $\theta|_{\partial U} = f|_{\partial U}$. Now, (a) above guarantees that $f \cong \theta$ in $D_{\partial U}(\overline{U}, E)$ and this together with $F \cong G$ in $A_{\partial U}(\overline{U}, E)$ (so $f \cong g$ in $D_{\partial U}(\overline{U}, E)$) and Remark 3 guarantees that $g \cong \theta$ in $D_{\partial U}(\overline{U}, E)$. Thus, (1) holds so Theorem 1 guarantees that F is essential in $A_{\partial U}(\overline{U}, E)$. A similar argument shows that, if F is essential in $A_{\partial U}(\overline{U}, E)$, then G is essential in $A_{\partial U}(\overline{U}, E)$. □

Theorem 3. *Let E be a Hausdorff locally convex topological vector space, U an open subset of E and $0 \in U$. Assume the zero map is in $\mathbf{A}(\overline{U}, E)$. Then, the zero map is essential in $A_{\partial U}(\overline{U}, E)$.*

Proof. Note $F(x) = \{0\}$ for $x \in \overline{U}$ (i.e., F is the zero map) and let $f \in D_{\partial U}(\overline{U}, E)$ be any selection of F. Note $f(x) = 0$ for $x \in \overline{U}$. Consider any map $g \in D_{\partial U}(\overline{U}, E)$ with $g|_{\partial U} = f|_{\partial U} = \{0\}$. We must show there exists a $x \in U$ with $x = g(x)$. Let

$$r(x) = \begin{cases} g(x), & x \in \overline{U}, \\ 0, & x \in E\backslash\overline{U}. \end{cases}$$

Note $r : E \to E$ is a continuous, compact map so [8] guarantees that there exists a $x \in E$ with $x = r(x)$. If $x \in E \backslash U$, then $r(x) = 0$, a contradiction since $0 \in U$. Thus, $x \in U$ and so $x = g(x)$. □

Now, we consider the above in the weak topology setting. Let X be a Hausdorff locally convex topological vector space and U a weakly open subset of C where C is a closed convex subset of X. Again, we consider a class $\mathbf{A}$ of maps.

Definition 8. *We say $f \in WD(\overline{U^w}, C)$ if $f : \overline{U^w} \to C$ is a weakly continuous, weakly compact map; here, $\overline{U^w}$ denotes the weak closure of U in C.*

Definition 9. *We say $f \in WD_{\partial U}(\overline{U^w}, C)$ if $f \in WD(\overline{U^w}, C)$ and $x \neq f(x)$ for $x \in \partial U$; here, ∂U denotes the weak boundary of U in C.*

Definition 10. *We say $F \in WA(\overline{U^w}, C)$ if $F : \overline{U^w} \to 2^C$ with $F \in \mathbf{A}(\overline{U^w}, C)$ and there exists a selection $f \in WD(\overline{U^w}, C)$ of F.*

Definition 11. *We say $F \in WA_{\partial U}(\overline{U^w}, C)$ if $F \in WA(\overline{U^w}, C)$ and $x \notin F(x)$ for $x \in \partial U$.*

Definition 12. *We say $F \in WA_{\partial U}(\overline{U^w}, C)$ is essential in $WA_{\partial U}(\overline{U^w}, C)$ if for any selection $f \in WD(\overline{U^w}, C)$ of F and any map $g \in WD_{\partial U}(\overline{U^w}, C)$ with $f|_{\partial U} = g|_{\partial U}$ there exists a $x \in U$ with $x = g(x)$.*

Definition 13. *Let $f, g \in WD_{\partial U}(\overline{U^w}, C)$. We say $f \cong g$ in $WD_{\partial U}(\overline{U^w}, C)$ if there exists a weakly continuous, weakly compact map $h : \overline{U^w} \times [0,1] \to C$ with $x \neq h_t(x)$ for any $x \in \partial U$ and $t \in (0,1)$ (here $h_t(x) = h(x,t)$), $h_0 = f$ and $h_1 = g$.*

Definition 14. *Let $F, G \in WA_{\partial U}(\overline{U^w}, C)$. We say $F \cong G$ in $WA_{\partial U}(\overline{U^w}, C)$ if for any selection $f \in WD_{\partial U}(\overline{U^w}, C)$ (respectively, $g \in WD_{\partial U}(\overline{U^w}, C)$) of F (respectively, of G) we have $f \cong g$ in $WD_{\partial U}(\overline{U^w}, C)$.*

Theorem 4. *Let X be a Hausdorff locally convex topological vector space and U a weakly open subset of C, where C is a closed convex subset of X. Suppose $F \in WA_{\partial U}(\overline{U^w}, C)$ and $G \in WA_{\partial U}(\overline{U^w}, C)$ is essential in $WA_{\partial U}(\overline{U^w}, C)$ and*

$$\begin{cases} \text{for any selection } f \in WD_{\partial U}(\overline{U^w}, C) \text{ (respectively, } g \in WD_{\partial U}(\overline{U^w}, C)) \\ \text{of } F \text{ (respectively, of } G) \text{ and any map } \theta \in WD_{\partial U}(\overline{U^w}, C) \\ \text{with } \theta|_{\partial U} = f|_{\partial U} \text{ we have } g \cong \theta \text{ in } WD_{\partial U}(\overline{U^w}, C). \end{cases} \tag{2}$$

Then, F is essential in $WA_{\partial U}(\overline{U^w}, C)$.

Proof. Let $f \in WD_{\partial U}(\overline{U^w}, C)$ be any selection of F and consider any map $\theta \in WD_{\partial U}(\overline{U^w}, C)$ with $\theta|_{\partial U} = f|_{\partial U}$. Let $g \in WD_{\partial U}(\overline{U^w}, C)$ be any selection of G. Now, (2) guarantees that there exists a weakly continuous, weakly compact map $h : \overline{U^w} \times [0,1] \to C$ with $x \neq h_t(x)$ for any $x \in \partial U$ and $t \in (0,1)$ (here $h_t(x) = h(x,t)$), $h_0 = g$ and $h_1 = \theta$. Let

$$\Omega = \left\{ x \in \overline{U^w} : \; x = h(x,t) \; \text{ for some } \; t \in [0,1] \right\}.$$

Recall that $X = (X, w)$, the space X endowed with the weak topology, is completely regular. Now, $\Omega \neq \emptyset$ is weakly closed and is in fact weakly compact with $\Omega \cap \partial U = \emptyset$. Thus, there exists a weakly continuous map $\mu : \overline{U^w} \to [0,1]$ with $\mu(\partial U) = 0$ and $\mu(\Omega) = 1$. Define the map r by $r(x) = h(x, \mu(x))$ and note $r \in WD_{\partial U}(\overline{U^w}, C)$ with $r|_{\partial U} = g|_{\partial U}$. Since G is essential in $WA_{\partial U}(\overline{U^w}, C)$, there exists a $x \in U$ with $x = r(x)$. Thus, $x \in \Omega$ so $x = h_1(x) = \theta(x)$. □

An obvious modification of the argument in Theorem 2 immediately yields the following result.

Theorem 5. *Let X be a Hausdorff locally convex topological vector space and U a weakly open subset of C, where C is a closed convex subset of X. Suppose F and G are two maps in $WA_{\partial U}(\overline{U}, C)$ with $F \cong G$ in $WA_{\partial U}(\overline{U}, C)$. Now, F is essential in $WA_{\partial U}(\overline{U}, C)$ if and only if G is essential in $WA_{\partial U}(\overline{U}, C)$.*

Now, we consider a generalization of essential maps, namely the d–essential maps [2]. Let E be a completely regular topological space and U an open subset of E. For any map $f \in D(\overline{U}, E)$, let $f^\star = I \times f : \overline{U} \to \overline{U} \times E$, with $I : \overline{U} \to \overline{U}$ given by $I(x) = x$, and let

$$d : \left\{ \left(f^\star\right)^{-1}(B) \right\} \cup \{\emptyset\} \to K \tag{3}$$

be any map with values in the nonempty set K; here, $B = \{(x,x) : \; x \in \overline{U}\}$.

Definition 15. *Let $F \in A_{\partial U}(\overline{U}, E)$ with $F^\star = I \times F$. We say $F^\star : \overline{U} \to 2^{\overline{U} \times E}$ is d–essential if, for any selection $f \in D(\overline{U}, E)$ of F and any map $g \in D_{\partial U}(\overline{U}, E)$ with $f|_{\partial U} = g|_{\partial U}$, we have that $d\left(\left(f^\star\right)^{-1}(B) \right) = d\left(\left(g^\star\right)^{-1}(B) \right) \neq d(\emptyset)$; here, $f^\star = I \times f$ and $g^\star = I \times g$.*

Remark 4. *If $F^\star$ is d–essential, then, for any selection $f \in D(\overline{U}, E)$ of F (with $f^\star = I \times f$), we have*

$$\emptyset \neq \left(f^\star\right)^{-1}(B) = \{x \in \overline{U} : \; (x, f(x)) \in B\},$$

so there exists a $x \in U$ with $x = f(x)$ (so, in particular, $x \in F(x)$).

Theorem 6. *Let E be a completely regular topological space, U an open subset of E, $B = \{(x,x) : \; x \in \overline{U}\}$, d is defined in(3), $F \in A_{\partial U}(\overline{U}, E)$, $G \in A_{\partial U}(\overline{U}, E)$ with $F^\star = I \times F$ and $G^\star = I \times G$. Suppose $G^\star$ is d–essential and*

$$\begin{cases} \text{for any selection } \; f \in D_{\partial U}(\overline{U}, E) \; \text{ (respectively, } \; g \in D_{\partial U}(\overline{U}, E)) \\ \text{of } \; F \; \text{ (respectively, of } \; G) \text{ and any map } \; \theta \in D_{\partial U}(\overline{U}, E) \\ \text{with } \; \theta|_{\partial U} = f|_{\partial U} \; \text{ we have } \; g \cong \theta \; \text{ in } \; D_{\partial U}(\overline{U}, E) \; \text{ and} \\ d\left(\left(f^\star\right)^{-1}(B) \right) = d\left(\left(g^\star\right)^{-1}(B) \right); \; \text{here } \; f^\star = I \times f \; \text{ and } \; g^\star = I \times g. \end{cases} \tag{4}$$

Then, $F^\star$ is d–essential.

Proof. Let $f \in D_{\partial U}(\overline{U}, E)$ be any selection of F and consider any map $\theta \in D_{\partial U}(\overline{U}, E)$ with $\theta|_{\partial U} = f|_{\partial U}$. We must show $d\left((f^\star)^{-1}(B)\right) = d\left((\theta^\star)^{-1}(B)\right) \neq d(\emptyset)$; here, $f^\star = I \times f$ and $\theta^\star = I \times \theta$. Let $g \in D_{\partial U}(\overline{U}, E)$ be any selection of G. Now, (4) guarantees that there exists a continuous, compact map $h : \overline{U} \times [0,1] \to E$ with $x \neq h_t(x)$ for any $x \in \partial U$ and $t \in (0,1)$ (here $h_t(x) = h(x,t)$), $h_0 = g$, $h_1 = \theta$ and $d\left((f^\star)^{-1}(B)\right) = d\left((g^\star)^{-1}(B)\right)$; here, $g^\star = I \times g$. Let $h^\star : \overline{U} \times [0,1] \to \overline{U} \times E$ be given by $h^\star(x,t) = (x, h(x,t))$ and let

$$\Omega = \left\{x \in \overline{U} :\ h^\star(x,t) \in B \ \text{ for some } \ t \in [0,1]\right\}.$$

Now, $\Omega \neq \emptyset$ is closed, compact and $\Omega \cap \partial U = \emptyset$ so there exists a continuous map $\mu : \overline{U} \to [0,1]$ with $\mu(\partial U) = 0$ and $\mu(\Omega) = 1$. Define the map r by $r(x) = h(x, \mu(x))$ and $r^\star = I \times r$. Now, $r \in D_{\partial U}(\overline{U}, E)$ with $r|_{\partial U} = g|_{\partial U}$. Since $G^\star$ is d–essential, then

$$d\left((g^\star)^{-1}(B)\right) = d\left((r^\star)^{-1}(B)\right) \neq d(\emptyset). \tag{5}$$

Now, since $\mu(\Omega) = 1$, we have

$$\begin{aligned} (r^\star)^{-1}(B) &= \left\{x \in \overline{U} :\ (x, h(x, \mu(x))) \in B\right\} = \left\{x \in \overline{U} :\ (x, h(x,1)) \in B\right\} \\ &= (\theta^\star)^{-1}(B), \end{aligned}$$

so, from the above and Equation (5), we have $d\left((f^\star)^{-1}(B)\right) = d\left((\theta^\star)^{-1}(B)\right) \neq d(\emptyset)$. □

Theorem 7. *Let E be a completely regular topological space, U an open subset of E, $B = \left\{(x,x) :\ x \in \overline{U}\right\}$ and d is defined in (3). Suppose F and G are two maps in $A_{\partial U}(\overline{U}, E)$ with $F^\star = I \times F$, $G^\star = I \times G$ and $F \cong G$ in $A_{\partial U}(\overline{U}, E)$. Then, $F^\star$ is d–essential if and only if $G^\star$ is d–essential.*

Proof. Assume $G^\star$ is d–essential. Let $f \in D_{\partial U}(\overline{U}, E)$ be any selection of F, $g \in D_{\partial U}(\overline{U}, E)$ be any selection of G and consider any map $\theta \in D_{\partial U}(\overline{U}, E)$ with $\theta|_{\partial U} = f|_{\partial U}$. If we show (4), then $F^\star$ is d–essential from Theorem 6. Now, $f \cong \theta$ in $D_{\partial U}(\overline{U}, E)$ together with $F \cong G$ in $A_{\partial U}(\overline{U}, E)$ (so $f \cong g$ in $D_{\partial U}(\overline{U}, E)$) guarantees that $g \cong \theta$ in $D_{\partial U}(\overline{U}, E)$. To complete (4), we need to show $d\left((f^\star)^{-1}(B)\right) = d\left((g^\star)^{-1}(B)\right)$; here, $f^\star = I \times f$ and $g^\star = I \times g$. We will show this by following the argument in Theorem 6. Note $G \cong F$ in $A_{\partial U}(\overline{U}, E)$ and let $h : \overline{U} \times [0,1] \to E$ be a continuous, compact map with $x \neq h_t(x)$ for any $x \in \partial U$ and $t \in (0,1)$ (here $h_t(x) = h(x,t)$), $h_0 = g$ and $h_1 = f$. Let $h^\star : \overline{U} \times [0,1] \to \overline{U} \times E$ be given by $h^\star(x,t) = (x, h(x,t))$ and let

$$\Omega = \left\{x \in \overline{U} :\ h^\star(x,t) \in B \ \text{ for some } \ t \in [0,1]\right\}.$$

Now, $\Omega \neq \emptyset$ and there exists a continuous map $\mu : \overline{U} \to [0,1]$ with $\mu(\partial U) = 0$ and $\mu(\Omega) = 1$. Define the map r by $r(x) = h(x, \mu(x))$ and $r^\star = I \times r$. Now, $r \in D_{\partial U}(\overline{U}, E)$ with $r|_{\partial U} = g|_{\partial U}$ so, since $G^\star$ is d–essential, then $d\left((g^\star)^{-1}(B)\right) = d\left((r^\star)^{-1}(B)\right) \neq d(\emptyset)$. Now, since $\mu(\Omega) = 1$, we have (see the argument in Theorem 6) $(r^\star)^{-1}(B) = (f^\star)^{-1}(B)$ and, as a result, we have $d\left((f^\star)^{-1}(B)\right) = d\left((g^\star)^{-1}(B)\right)$. □

Remark 5. *It is also easy to extend the above ideas to other natural situations. Let E be a (Hausdorff) topological vector space (so automatically completely regular), Y a topological vector space, and U an open subset of E. In addition, let $L : dom\, L \subseteq E \to Y$ be a linear (not necessarily continuous) single valued map; here, $dom\, L$ is a vector subspace of E. Finally, $T : E \to Y$ will be a linear, continuous single valued map with $L + T : dom\, L \to Y$ an isomorphism (i.e., a linear homeomorphism); for convenience we say $T \in H_L(E, Y)$.*

We say $F \in A(\overline{U}, Y; L, T)$ *if* $(L+T)^{-1}(F+T) \in A(\overline{U}, E)$ *and we could discuss essential and d–essential in this situation.*

Now, we present an example to illustrate our theory.

Example 1. *Let* E *be a Hausdorff locally convex topological vector space,* U *an open subset of* E, $0 \in U$ *and* $\overline{U}$ *paracompact. In this case, we say that* $F \in \mathbf{A}(\overline{U}, E)$ *if* $F \in PK(\overline{U}, E)$ *(see Remark 1) is a compact map. Let* $F \in A_{\partial U}(\overline{U}, E)$ *and assume* $x \notin \lambda F(x)$ *for* $x \in \partial U$ *and* $\lambda \in (0,1)$*. Then,* $F \cong 0$ *in* $A_{\partial U}(\overline{U}, E)$*. To see this, let* $f \in D_{\partial U}(\overline{U}, E)$ *be any selection of* F *and let* $h : \overline{U} \times [0,1]$ *be given by* $h(x,t) = t\,f(x)$*. Note that* $h_0 = 0$*,* $h_1 = f$ *and* $x \notin h_t(x)$ *for* $x \in \partial U$ *and* $\lambda \in (0,1)$ *so* $f \cong 0$ *in* $D_{\partial U}(\overline{U}, E)$*. Now, Theorems 2 and 3 guarantee that* F *is essential in* $A_{\partial U}(\overline{U}, E)$*.*

3. Conclusions

In this paper, we prove that, for two set-valued maps F and G with continuous compact selections and $F \cong G$, then one being essential (or *d*–essential) guarantees that the other is essential (or *d*–essential).

Funding: This research received no external funding.

Conflicts of Interest: The author declares no conflict of interest.

References

1. Granas, A. Sur la méthode de continuité de Poincaré. *C.R. Acad. Sci. Paris* **1976**, 282, 983–985.
2. Precup, R. On the topological transversality principle. *Nonlinear Anal.* **1993**, *20*, 1–9. [CrossRef]
3. Gabor, G.; Gorniewicz, L.; Slosarski, M. Generalized topological essentiality and coincidence points of multivalued maps. *Set-Valued Anal.* **2009**, *17*, 1–19. [CrossRef]
4. O'Regan, D. Essential maps and coincidence principles for general classes of maps. *Filomat* **2017**, *31*, 3553–3558. [CrossRef]
5. O'Regan, D. Topological transversality principles and general coincidence theory. *An. Stiint. Univ. Ovidius Constanta Ser. Mat.* **2017**, *25*, 159–170. [CrossRef]
6. O'Regan, D. Topological fixed point theory for compact multifunctions via homotopy and essential maps. *Topol. Appl.* **2019**, *265*, 106819. [CrossRef]
7. Lin, L.J.; Park, S.; You, Z.T. Remarks on fixed points, maximal elements and equilibria of generalized games. *J. Math. Anal. Appl.* **1999**, *233*, 581–596. [CrossRef]
8. Himmelberg, C.J. Fixed points of compact multifunctions. *J. Math. Anal. Appl.* **1972**, *38*, 205–207. [CrossRef]

mathematics

MDPI

Article

A Class of Equations with Three Solutions

Biagio Ricceri

Department of Mathematics and Informatics, University of Catania, Viale A. Doria 6, 95125 Catania, Italy; ricceri@dmi.unict.it

Received: 5 March 2020; Accepted: 29 March 2020; Published: 1 April 2020

Abstract: Here is one of the results obtained in this paper: Let $\Omega \subset \mathbf{R}^n$ be a smooth bounded domain, let $q > 1$, with $q < \frac{n+2}{n-2}$ if $n \geq 3$ and let λ_1 be the first eigenvalue of the problem $-\Delta u = \lambda u$ in Ω, $u = 0$ on $\partial\Omega$. Then, for every $\lambda > \lambda_1$ and for every convex set $S \subseteq L^\infty(\Omega)$ dense in $L^2(\Omega)$, there exists $\alpha \in S$ such that the problem $-\Delta u = \lambda(u^+ - (u^+)^q) + \alpha(x)$ in Ω, $u = 0$ on $\partial\Omega$, has at least three weak solutions, two of which are global minima in $H_0^1(\Omega)$ of the functional $u \to \frac{1}{2}\int_\Omega |\nabla u(x)|^2 dx - \lambda \int_\Omega \left(\frac{1}{2}|u^+(x)|^2 - \frac{1}{q+1}|u^+(x)|^{q+1}\right) dx - \int_\Omega \alpha(x)u(x)dx$ where $u^+ = \max\{u, 0\}$.

Keywords: minimax; multiplicity; global minima

---

1. Introduction

There is no doubt that the study of nonlinear PDEs lies in the core of Nonlinear Analysis. In turn, one of the most studied topics concerning nonlinear PDEs is the multiplicity of solutions. On the other hand, the study of the global minima of integral functionals is essentially the central subject of the Calculus of Variations. In the light of these facts, it is hardly understable why the number of the known results on multiple global minima of integral functionals is extremely low. Certainly, this is not due to a lack of intrinsic mathematical interest. Probably, the reason could reside in the fact that there is not an abstract tool which has the same popularity as the one that, for instance, the Lyusternik–Schnirelmann theory and the Morse theory have in dealing with multiple solutions for nonlinear PDEs.

Abstract results on the multiplicity of global minima, however, are already present in the literature. We allude to the result first obtained in [1] and then extended in [2,3] which ensures the existence of at least two global minima provided that a strict minimax inequality holds. We already have obtained a variety of applications upon different ways of checking the required strict inequality ([4–6]).

The aim of the present paper is to establish an application of Theorem 1 of [7] which is itself an application of the main result in [3]. Precisely, we first establish a general result which ensures the existence of three solutions for a certain equation provided that another related one has no non-zero solutions (Theorem 1). Then, we present an application to nonlinear elliptic equations (Theorem 2).

2. Results

In the sequel, $(X, \|\cdot\|_X)$ is a reflexive real Banach space, $(Y, \langle\cdot,\cdot\rangle_Y)$ is a real Hilbert space, $I, \psi : X \to \mathbf{R}$ are two C^1 functionals, with $I(0) = \psi(0) = 0$ and $\sup_{\mathbf{R}} \psi > 0$, $\varphi : X \to Y$ is a C^1 operator, with $\varphi(0) = 0$. For each fixed $y \in Y$, we denote by $\partial_x\langle\varphi(\cdot), y\rangle$ the derivative of the functional $x \to \langle\varphi(x), y\rangle$. Clearly, one has

$$\partial_x\langle\varphi(x), y\rangle(u) = \langle\varphi'(x)(u), y\rangle$$

for all $x, u \in X$.

We say that I is coercive if $\lim_{\|x\|_X \to +\infty} I(x) = +\infty$. We also say that I' admits a continuous inverse on X^* if there exists a continuous operator $T : X^* \to X$ such that $T(I'(x)) = x$ for all $x \in X$.

Here is our abstract result:

Theorem 1. *Let I be weakly lower semicontinuous and coercive, and let I' admit a continuous inverse on X^*. Moreover, assume that the operators φ' and ψ' are compact and that*

$$\lim_{\|x\|_X\to+\infty}\frac{\langle\varphi(x),y\rangle_Y}{I(x)}=0 \tag{1}$$

for all y in a convex and dense set $V\subseteq Y$.

Set

$$\theta^* := \inf_{x\in\psi^{-1}(]0,+\infty[)}\frac{I(x)}{\psi(x)},$$

$$\tilde{\theta} := \begin{cases} \liminf_{x\in\psi^{-1}(]0,+\infty[),\|x\|_X\to+\infty}\frac{I(x)}{\psi(x)} & \text{if } \psi^{-1}(]0,+\infty[) \text{ is unbounded}\\ +\infty & \text{otherwise}\end{cases}$$

and assume that

$$\theta^* < \tilde{\theta}.$$

Then, for each $\lambda\in]\theta^*,\tilde{\theta}[$, with $\lambda\geq 0$, either the equation

$$I'(x) = -\partial_x\langle\varphi(x),\varphi(x)\rangle + \lambda\psi'(x)$$

has a non-zero solution, or, for each convex set $S\subseteq V$ dense in Y, there exists $\tilde{y}\in S$ such that the equation

$$I'(x) = \partial_x\langle\varphi(x),\tilde{y}\rangle_Y + \lambda\psi'(x)$$

has at least three solutions, two of which are global minima in X of the functional

$$x\to I(x) - \langle\varphi(x),\tilde{y}\rangle_Y - \lambda\psi(x).$$

As it was said in the Introduction, the main tool to prove Theorem 1 is a result recently obtained in [7]. For reader's convenience, we now recall its statement:

Theorem 2. *([7], Theorem 1). - Let X, E be two real reflexive Banach spaces and let $\Phi : X\times E\to\mathbf{R}$ be a C^1 functional satisfying the following conditions:*

(*a*) *the functional $\Phi(x,\cdot)$ is quasi-concave for all $x\in X$ and the functional $-\Phi(x_0,\cdot)$ is coercive for some $x_0\in X$;*

(*b*) *there exists a convex set $S\subseteq E$ dense in E, such that, for each $y\in S$, the functional $\Phi(\cdot,y)$ is weakly lower semicontinuous, coercive and satisfies the Palais-Smale condition.*

Then, either the system

$$\begin{cases}\Phi'_x(x,y)=0\\ \Phi'_y(x,y)=0\end{cases}$$

has a solution (x^*,y^*) such that

$$\Phi(x^*,y^*) = \inf_{x\in X}\Phi(x,y^*) = \sup_{y\in E}\Phi(x^*,y),$$

or, for every convex set $T\subseteq S$ dense in E, there exists $\tilde{y}\in T$ such that equation

$$\Phi'_x(x,\tilde{y})=0$$

has at least three solutions, two of which are global minima in X of the functional $\Phi(\cdot,\tilde{y})$.

Proof of Theorem 1. Fix $\lambda \in]\theta^*, \tilde{\theta}[$, with $\lambda \geq 0$. Assume that the equation

$$I'(x) = -\partial_x \langle \varphi(x), \varphi(x) \rangle + \lambda \psi'(x)$$

has no non-zero solution. Fix a convex set $S \subseteq Y$ dense in Y. We have to show that there exists $\tilde{y} \in S$ such that the equation

$$I'(x) = \partial_x \langle \varphi(x), \tilde{y} \rangle_Y + \lambda \psi'(x)$$

has at least three solutions, two of which are global minima in X of the functional $x \to I(x) - \langle \varphi(x), \tilde{y} \rangle_Y - \lambda \psi(x)$. To this end, let us apply Theorem 2. Consider the functional $\Phi : X \times Y \to \mathbf{R}$ defined by

$$\Phi(x,y) = I(x) - \frac{1}{2}\|y\|_Y^2 - \langle \varphi(x), y \rangle - \lambda \psi(x)$$

for all $(x,y) \in X \times Y$. Of course, Φ is C^1 and, for each $x \in X$, $\Phi(x,\cdot)$ is concave and $-\Phi(x,\cdot)$ is coercive. Fix $y \in Y$. Let us show that the operator $\partial_x \langle \varphi(\cdot), y \rangle$ is compact. To this end, let $\{x_n\}$ be a bounded sequence in X. Since φ' is compact, up to a subsequence, $\{\varphi'(x_n)\}$ converges in $\mathcal{L}(X,Y)$ to some η. That is

$$\lim_{n \to \infty} \sup_{\|u\|_X = 1} \|\varphi'(x_n)(u) - \eta(u)\|_Y = 0\,.$$

On the other hand, we have

$$\begin{aligned} \sup_{\|u\|_X=1} |\partial_x \langle \varphi(x_n), y \rangle(u) - \langle \eta(u), y \rangle| &= \sup_{\|u\|_X=1} |\langle \varphi'(x_n)(u), y \rangle - \langle \eta(u), y \rangle| \\ &\leq \sup_{\|u\|_X=1} \|\varphi'(x_n)(u) - \eta(u)\|_Y \|y\|_Y \end{aligned}$$

and so the sequence $\{\partial_x \langle \varphi(x_n), y \rangle(\cdot)\}$ converges in X^* to $\eta(\cdot)(y)$. Then, since ψ' is compact, the operator $\partial_x \langle \varphi(\cdot), y \rangle + \lambda \psi'(\cdot)$ is compact too. From this, it follows that $\langle \varphi(\cdot), y \rangle + \lambda \psi(\cdot)$ is sequentially weakly continuous ([8], Corollary 41.9). If $\|x\|_X$ is large enough, we have $I(x) > 0$ and so we can write

$$\Phi(x,y) = I(x) \left(1 - \frac{\frac{1}{2}\|y\|_Y^2 + \langle \varphi(x), y \rangle + \lambda \psi(x)}{I(x)}\right). \tag{2}$$

In view of (1), we also have

$$\liminf_{\|x\|_X \to +\infty} \left(1 - \frac{\frac{1}{2}\|y\|_Y^2 + \langle \varphi(x), y \rangle + \lambda \psi(x)}{I(x)}\right) = 1 - \limsup_{\|x\|_X \to +\infty} \frac{\lambda \psi(x)}{I(x)}\,. \tag{3}$$

We claim that

$$\limsup_{\|x\| \to +\infty} \frac{\lambda \psi(x)}{I(x)} < 1\,. \tag{4}$$

This is clear if either $\lambda = 0$ or $\limsup_{\|x\|_X \to +\infty} \frac{\psi(x)}{I(x)} \leq 0$. If $\lambda > 0$ and $\limsup_{\|x\|_X \to +\infty} \frac{\psi(x)}{I(x)} > 0$, then (4) is equivalent to

$$\limsup_{\|x\|_X \to +\infty} \frac{\psi(x)}{I(x)} < +\infty$$

and

$$\lambda < \frac{1}{\limsup_{\|x\|_X \to +\infty} \frac{\psi(x)}{I(x)}}\,. \tag{5}$$

But

$$\frac{1}{\limsup_{\|x\|_X \to +\infty} \frac{\psi(x)}{I(x)}} = \liminf_{x \in \psi^{-1}(]0,+\infty[), \|x\|_X \to +\infty} \frac{I(x)}{\psi(x)}\,,$$

and so (5) is satisfied just since $\lambda < \tilde{\theta}$. Since I is coercive and weakly lower semicontinuous, the functional $\Phi(\cdot, y)$ turns out to be coercive, in view of (2), (3), (4), and weakly lower semicontinuous, in view of the Eberlein-Smulyan theorem. Finally, since I' admits a continuous inverse on X^*, $\Phi(\cdot, y)$ satisfies the Palais-Smale condition in view of Example 38.25 of [8]. Hence, Φ satisfies the assumptions of Theorem 2. Now, we claim that there is no solution (x^*, y^*) of the system

$$\begin{cases} \Phi'_x(x,y) = 0 \\ \Phi'_y(x,y) = 0 \end{cases}$$

such that

$$\Phi(x^*, y^*) = \inf_{x \in X} \Phi(x, y^*) .$$

Arguing by contradiction, assume that such a (x^*, y^*) does exist. This amounts to say that

$$\begin{cases} I'(x^*) = \partial_x \langle \varphi(x^*), y^* \rangle + \lambda \psi'(x^*) \\ y^* = -\varphi(x^*) \end{cases}$$

and

$$I(x^*) - \langle \varphi(x^*), y^* \rangle - \lambda \psi(x^*) = \inf_{x \in X} (I(x) - \langle \varphi(x), y^* \rangle - \lambda \psi(x)) . \tag{6}$$

Therefore

$$I'(x^*) = -\partial_x \langle \varphi(x^*), \varphi(x^*) \rangle + \lambda \psi'(x^*) .$$

So, by the initial assumption, we have $x^* = 0$ and hence $y^* = 0$ (recall that $\varphi(0) = 0$). As a consequence, since $I(0) = \psi(0) = 0$, (6) becomes

$$\inf_{x \in X} (I(x) - \lambda \psi(x)) = 0 . \tag{7}$$

Now, notice that (7) contradicts the fact that $\lambda > \theta^*$. Hence, *a fortiori*, the system

$$\begin{cases} \Phi'_x(x,y) = 0 \\ \Phi'_y(x,y) = 0 \end{cases}$$

has no solution (x^*, y^*) such that

$$\Phi(x^*, y^*) = \inf_{x \in X} \Phi(x, y^*) = \sup_{y \in Y} \Phi(x^*, y)$$

and then the existence of $\tilde{y} \in S$ is directly ensured by Theorem 2. □

We now present an application of Theorem 1 to a class of nonlinear elliptic equations. Let $\Omega \subset \mathbf{R}^n$ be a smooth bounded domain. We denote by $\mathcal{A}$ the class of all Carathéodory's functions $f : \Omega \times \mathbf{R} \to \mathbf{R}$ such that, for each $u, v \in H^1_0(\Omega)$, the function $x \to f(x, u(x))v(x)$ lies in $L^1(\Omega)$. For $f \in \mathcal{A}$, we consider the Dirichlet problem

$$\begin{cases} -\Delta u = f(x,u) & \text{in } \Omega \\ u = 0 & \text{on } \partial\Omega . \end{cases}$$

As usual, a weak solution of the problem is any $u \in H^1_0(\Omega)$ such that

$$\int_\Omega \nabla u(x) \nabla v(x) dx = \int_\Omega f(x, u(x)) v(x) dx$$

for all $v \in H^1_0(\Omega)$.

Also, we denote by λ_1 the first eigenvalue of the Dirichlet problem

$$\begin{cases} -\Delta u = \lambda u & \text{in } \Omega \\ u = 0 & \text{on } \partial\Omega \,. \end{cases}$$

For any continuous function $f : \mathbf{R} \to \mathbf{R}$, we set $F(\xi) = \int_0^\xi f(t)dt$ for all $\xi \in \mathbf{R}$.

Theorem 3. *Let $f, g : \mathbf{R} \to \mathbf{R}$ be two continuous functions satisfying the following growth conditions:*

(a) *if $n \leq 3$, one has*

$$\lim_{|\xi| \to +\infty} \frac{|F(\xi)|}{\xi^2} = 0\,;$$

(b) *if $n \geq 2$, there exist $p, q > 0$, with $p < \frac{2}{n-2}$, $q < \frac{n+2}{n-2}$ if $n \geq 3$, such that*

$$\sup_{\xi \in \mathbf{R}} \frac{|f(\xi)|}{1 + |\xi|^p} < +\infty\,,$$

$$\sup_{\xi \in \mathbf{R}} \frac{|g(\xi)|}{1 + |\xi|^q} < +\infty\,.$$

Set

$$\rho := \limsup_{|\xi| \to +\infty} \frac{G(\xi)}{\xi^2}\,,$$

$$\sigma := \max\left\{ \liminf_{\xi \to 0^+} \frac{G(\xi)}{\xi^2}, \liminf_{\xi \to 0^-} \frac{G(\xi)}{\xi^2} \right\}$$

and assume that

$$\max\{\rho, 0\} < \sigma\,.$$

Then, for every $\lambda \in \left] \frac{\lambda_1}{2\sigma}, \frac{\lambda_1}{2\max\{\rho,0\}} \right[$ (with the conventions $\frac{\lambda_1}{+\infty} = 0$, $\frac{\lambda_1}{0} = +\infty$), either the problem

$$\begin{cases} -\Delta u = -F(u)f(u) + \lambda g(u) & \text{in } \Omega \\ u = 0 & \text{on } \partial\Omega \end{cases} \tag{8}$$

has a non-zero weak solution, or, for every convex set $S \subseteq L^\infty(\Omega)$ dense in $L^2(\Omega)$, there exists $\alpha \in S$ such that the problem

$$\begin{cases} -\Delta u = \alpha(x)f(u) + \lambda g(u) & \text{in } \Omega \\ u = 0 & \text{on } \partial\Omega \end{cases} \tag{9}$$

has at least three weak solutions, two of which are global minima in $H_0^1(\Omega)$ of the functional

$$u \to \frac{1}{2}\int_\Omega |\nabla u(x)|^2 dx - \int_\Omega \alpha(x)F(u(x))dx - \lambda \int_\Omega G(u(x))dx\,.$$

Proof. We are going to apply Theorem 1 taking $X = H_0^1(\Omega)$, $Y = L^2(\Omega)$, with their usual scalar products (that is, $\langle u, v\rangle_X = \int_\Omega \nabla u(x) \nabla v(x) dx$ and $\langle u, v\rangle_Y = \int_\Omega u(x)v(x)dx$), $V = L^\infty(\Omega)$ and

$$I(u) = \frac{1}{2}\|u\|_X^2\,,$$

$$\varphi(u) = F \circ u\,,$$

$$\psi(u) = \int_\Omega G(u(x))dx$$

for all $u \in X$. In view of (b), thanks to the Sobolev embedding theorem, the operator φ and the functional ψ are C^1, with compact derivative. Moreover, the solutions of the equation

$$I'(u) = -\partial_u \langle \varphi(u), \varphi(u) \rangle_Y + \lambda \psi'(u)$$

are weak solutions of (8) and, for each $\alpha \in Y$, the solutions of the equation

$$I'(u) = \partial_u \langle \varphi(u), \alpha \rangle_Y + \lambda \psi'(u)$$

are weak solutions of (9). Moreover, condition (1) follows readily from (a) which is automatically satisfied when $n \geq 4$ since $p < \frac{2}{n-2}$. We claim that

$$\limsup_{\|u\|_X \to +\infty} \frac{\psi(u)}{\|u\|_X^2} \leq \frac{\rho}{\lambda_1} . \tag{10}$$

Indeed, fix $\nu > \rho$. Then, there exists $\delta > 0$ such that

$$G(\xi) \leq \nu \xi^2 \tag{11}$$

for all $x \in \mathbf{R} \setminus [-\delta, \delta]$. Fix $u \in X \setminus \{0\}$. From (11) we clearly obtain

$$\psi(u) \leq \nu \|u\|_Y^2 + \text{meas}(\Omega) \sup_{[-\delta,\delta]} G \leq \nu \frac{\|u\|_X^2}{\lambda_1} + \text{meas}(\Omega) \sup_{[-\delta,\delta]} G$$

and so

$$\limsup_{\|u\|_X \to +\infty} \frac{\psi(u)}{\|u\|_X^2} \leq \frac{\nu}{\lambda_1} . \tag{12}$$

Now, we get (10) passing in (12) to the limit for ν tending to ρ. We also claim that

$$\frac{\sigma}{\lambda_1} \leq \sup_{u \in X \setminus \{0\}} \frac{\psi(u)}{\|u\|_X^2} . \tag{13}$$

Indeed, fix $\eta < \sigma$. For instance, let $\sigma = \liminf_{\xi \to 0^+} \frac{G(\xi)}{\xi^2}$. Then, there exists $\eta > 0$ such that

$$G(\xi) \geq \eta \xi^2 \tag{14}$$

for all $\xi \in [0, \eta]$. Fix any $v \in H_0^1(\Omega)$ such that $\|v\|_X^2 = \lambda_1 \|v\|_Y^2$ and $v(\Omega) \subseteq [0, \eta]$. From (14) we obtain

$$\psi(v) \geq \eta \|v\|_Y^2$$

and so

$$\sup_{u \in X \setminus \{0\}} \frac{\psi(u)}{\|u\|_X^2} \geq \frac{\psi(v)}{\|v\|_X^2} \geq \frac{\eta}{\lambda_1} . \tag{15}$$

Now, (13) is obtained from (15) passing to the limit for η tending to σ. Now, fix $\lambda \in \left] \frac{\lambda_1}{2\sigma}, \frac{\lambda_1}{2\max\{\rho,0\}} \right[$. Then, from (10) and (13), we obtain

$$\limsup_{\|u\|_X \to +\infty} \frac{\psi(u)}{I(u)} < \frac{1}{\lambda} < \sup_{u \in X \setminus \{0\}} \frac{\psi(u)}{I(u)} .$$

This readily implies that $\theta^* < \lambda < \tilde{\theta}$ and the conclusion is directly provided by Theorem 1. $\square$

Corollary 1. *Let the assumptions of Theorem 3 be satisfied and let* $\lambda \in \left]\frac{\lambda_1}{2\sigma}, \frac{\lambda_1}{2\max\{\rho,0\}}\right[$ *satisfy*

$$\sup_{\xi \in \mathbf{R}}(\lambda g(\xi) - F(\xi)f(\xi))\xi \leq 0 . \tag{16}$$

Then, for every convex set $S \subseteq L^\infty(\Omega)$ *dense in* $L^2(\Omega)$*, there exists* $\alpha \in S$ *such that the problem*

$$\begin{cases} -\Delta u = \alpha(x)f(u) + \lambda g(u) & \text{in } \Omega \\ u = 0 & \text{on } \partial\Omega \end{cases}$$

has at least three weak solutions, two of which are global minima in $H_0^1(\Omega)$ *of the functional*

$$u \to \frac{1}{2}\int_\Omega |\nabla u(x)|^2 dx - \int_\Omega \alpha(x)F(u(x))dx - \lambda \int_\Omega G(u(x))dx .$$

Proof. It suffices to observe that, in view of (16), 0 is the only solution of (8) and then to apply Theorem 3. □

Finally, notice the following remarkable corollary of Corollary 1:

Corollary 2. *Let* $q > 1$*, with* $q < \frac{n+2}{n-2}$ *if* $n \geq 3$*. Let* $h : \mathbf{R} \to \mathbf{R}$ *be a non-negative continuous function, with* $\inf_{[0,1]} h > 0$*, satisfying conditions (a) and (b) of Theorem 3 for* $f = h$*.*

Then, for every $\lambda > \lambda_1$ *and for every convex set* $S \subseteq L^\infty(\Omega)$ *dense in* $L^2(\Omega)$*, there exists* $\alpha \in S$ *such that the problem*

$$\begin{cases} -\Delta u = \alpha(x)h(u) + \lambda(u^+ - (u^+)^q) & \text{in } \Omega \\ u = 0 & \text{on } \partial\Omega \end{cases}$$

has at least three weak solutions, two of which are global minima in $H_0^1(\Omega)$ *of the functional*

$$u \to \frac{1}{2}\int_\Omega |\nabla u(x)|^2 dx - \int_\Omega \alpha(x)H(u(x))dx - \lambda \int_\Omega \left(\frac{1}{2}|u^+(x)|^2 - \frac{1}{q+1}|u^+(x)|^{q+1}\right) dx .$$

Proof. Fix $\lambda > \lambda_1$. Notice that, since $\inf_{[0,1]} h > 0$, the number

$$\gamma := \inf_{\xi \in]0,1]} \frac{H(\xi)h(\xi)}{\xi}$$

is positive. Now, we are going to apply Corollary 1 taking

$$f(\xi) = \sqrt{\frac{\lambda}{\gamma}}h(\xi)$$

and

$$g(\xi) = \xi^+ - (\xi^+)^q .$$

Of course (with the notations of Theorem 3), $\rho = 0$ and $\sigma = \frac{1}{2}$. Since f in non-negative, Ff is so in $[0, +\infty[$ and non-positive in $]-\infty, 0]$. Therefore, (16) is satisfied for all $\xi \in \mathbf{R} \setminus [0,1]$ since g has the opposite sign of Ff in that set. Now, let $\xi \in]0,1]$. We have

$$\frac{F(\xi)f(\xi)}{\xi} = \frac{\lambda}{\gamma}\frac{H(\xi)h(\xi)}{\xi} \geq \lambda \geq \lambda(1 - \xi^{q-1})$$

which gives (16). Now, let $S \subseteq L^\infty(\Omega)$ be any convex set dense in $L^2(\Omega)$. Then, the set $\sqrt{\frac{\gamma}{\lambda}}S$ is convex and dense in $L^2(\Omega)$ and the conclusion follows applying Corollary 1 with this set. □

Remark 1. *We are not aware of known results close enough to Theorems 1 and 3 in order to do a proper comparison. We refer to the monographs [9,10] for an account on multiplicity results for nonlinear PDEs.*

Funding: This research received no external funding.

Acknowledgments: The author has been supported by the Gruppo Nazionale per l'Analisi Matematica, la Probabilità e le loro Applicazioni (GNAMPA) of the Istituto Nazionale di Alta Matematica (INdAM) and by the Università degli Studi di Catania, "Piano della Ricerca 2016/2018 Linea di intervento 2".

Conflicts of Interest: The author declares no conflict of interest.

References

1. Ricceri, B. Multiplicity of global minima for parametrized functions. *Rend. Lincei Mat. Appl.* **2010**, *21*, 47–57. [CrossRef]
2. Ricceri, B. A strict minimax inequality criterion and some of its consequences. *Positivity* **2012**, *16*, 455–470. [CrossRef]
3. Ricceri, B. On a minimax theorem: An improvement, a new proof and an overview of its applications. *Minimax Theory Appl.* **2017**, *2*, 99–152.
4. Ricceri, B. A range property related to non-expansive operators. *Mathematika* **2014**, *60*, 232–236. [CrossRef]
5. Ricceri, B. Singular points of non-monotone potential operators. *J. Nonlinear Convex Anal.* **2015**, *16*, 1123–1129.
6. Ricceri, B. Miscellaneous applications of certain minimax theorems II. *Acta Math. Vietnam.* **2020**, in press.
7. Ricceri, B. An alternative theorem for gradient systems. *arXiv* **2020**, arXiv:2002.01413. Available online: https://arxiv.org/abs/2002.01413 (accessed on 4 February 2020).
8. Zeidler, E. *Nonlinear Functional Analysis and Its Applications*; Springer: New York, NY, USA, 1985; Volume III.
9. Kristály, A.; Rădulescu, V.D.; Varga, C. *Variational Principles in Mathematical Physics, Geometry, and Economics: Qualitative Analysis of Nonlinear Equations and Unilateral Problems*; Cambridge University Press: Cambridge, UK, 2010.
10. Molica Bisci, G.M.; Radulescu, V.D.; Servadei, R. *Variational Methods for Nonlocal Fractional Problems*; Cambridge University Press: Cambridge, UK, 2016.

mathematics

Correction

Correction: Ricceri, B. A Class of Equations with Three Solutions. *Mathematics* 2020, *8*, 478

Biagio Ricceri

Department of Mathematics and Informatics, University of Catania, Viale A. Doria 6, 95125 Catania, Italy; ricceri@dmi.unict.it

Received: 15 October 2020; Accepted: 22 December 2020; Published: 5 January 2021

The author wishes to make the following correction to this paper [1]:

Everywhere it occurs, the phrase "for every convex set $S \subseteq H_0^1(\Omega)$ dense in $H_0^1(\Omega)$" should be replaced with "for every convex set $S \subseteq L^\infty(\Omega)$ dense in $L^2(\Omega)$".

Actually, thanks to (b) of Theorem 2, condition (1) can be weakened to

$$\lim_{\|x\|_X \to +\infty} \frac{\langle \varphi(x), y \rangle_Y}{I(x)} = 0 \qquad (1)$$

for all y in a convex and dense set $V \subseteq Y$. Then, in the conclusion of Theorem 1, we can replace "$S \subseteq Y$" with "$S \subseteq V$". Finally, in the proof of Theorem 3, we take $V = L^\infty(\Omega)$, so that condition (a) is actually enough to prove equality (1).

The author would like to apologize for any inconvenience caused to the readers by these changes. The changes do not affect the scientific results. The original article has been updated.

Conflicts of Interest: The author declare no conflict of interest.

Reference

1. Ricceri, B. A Class of Equations with Three Solutions. *Mathematics* **2020**, *8*, 478. [CrossRef]

mathematics
MDPI

Article

Positive Solutions for Discontinuous Systems via a Multivalued Vector Version of Krasnosel'skiĭ's Fixed Point Theorem in Cones

Rodrigo López Pouso [1], Radu Precup [2,*] and Jorge Rodríguez-López [1]

1 Departamento de Estatística, Análise Matemática e Optimización, Instituto de Matemáticas, Universidade de Santiago de Compostela, Facultade de Matemáticas, Campus Vida, 15782 Santiago, Spain; rodrigo.lopez@usc.es (R.L.P.); jorgerodriguez.lopez@usc.es (J.R.-L.)
2 Department of Mathematics, Babeş-Bolyai University, 400084 Cluj, Romania
* Correspondence: r.precup@math.ubbcluj.ro

Received: 12 February 2019; Accepted: 14 May 2019; Published: 20 May 2019

Abstract: We establish the existence of positive solutions for systems of second–order differential equations with discontinuous nonlinear terms. To this aim, we give a multivalued vector version of Krasnosel'skiĭ's fixed point theorem in cones which we apply to a regularization of the discontinuous integral operator associated to the differential system. We include several examples to illustrate our theory.

Keywords: Krasnosel'skiĭ's fixed point theorem; positive solutions; discontinuous differential equations; differential system

1. Introduction

We study the existence and localization of positive solutions for the system

$$\begin{cases} u_1''(t) + g_1(t)f_1(t, u_1(t), u_2(t)) = 0, \\ u_2''(t) + g_2(t)f_2(t, u_1(t), u_2(t)) = 0, \end{cases}$$

subject to the Sturm–Liouville boundary conditions (7).

The novelties in this paper are in two directions. On the one hand, we allow the functions f_i $(i = 1, 2)$ to be discontinuous with respect to the unknown over some time-dependent sets, see Definitions 1 and 2. On the other hand, in order to localize the solutions of the system, we shall establish a multivalued vector version of Krasnosel'skiĭ's fixed point theorem which allows different asymptotic behaviors in the nonlinearities f_1 and f_2, see Remark 3.

The existence of discontinuities in the functions f_1 or f_2 makes impossible to apply directly the standard fixed point theorems in cones for compact operators since the integral operator corresponding to the differential problem is not necessarily continuous. In order to avoid this difficulty, we regularize the possibly discontinuous operator obtaining an upper semicontinuous multivalued one. Then we look for fixed points of this multivalued mapping that are proved to be Carathéodory solutions for the differential system. In the case of scalar problems, similar ideas appear in the papers [1–3].

This approach of using set-valued analysis in the study of discontinuous problems is a classical one, see [4]. Nevertheless, the regularization is usually made in the nonlinearities transforming the problem into a differential inclusion and the solutions are often given in the sense of the set-valued analysis (Krasovskij and Filippov solutions [5,6]), see e.g., [7,8]. Similar ideas are also used in the papers [5,9] where there are provided some sufficient conditions for the Krasovskij solutions to be Carathéodory solutions. Recently, second-order scalar discontinuous problems have been investigated by using variational methods [10–12]. However, in these papers there are not considered time-dependent discontinuity sets. Observe also that a lot of existence results for discontinuous differential problems are based on monotonicity hypotheses on their nonlinear parts, see [13], but such assumptions are not necessary in our approach.

Going from scalar discontinuous problems to systems of discontinuous equations is not trivial and it makes possible to consider two different notions for the discontinuity sets. The first approach (see Definition 1 and Theorem 3) allows to study the discontinuities in each variable independently. For instance, it guarantees the existence of a positive solution for the following particular system

$$\begin{cases} -x''(t) = x^2 + x^2y^2H(1-x)H(1-y), \\ -y''(t) = \sqrt{x} + \sqrt{y} + H(x-1)H(y-1), \end{cases}$$

subject to the Sturm–Liouville boundary conditions, where $H : \mathbb{R} \to \mathbb{R}$ is the Heaviside step function given by

$$H(x) = \begin{cases} 0, & \text{if } x \le 0, \\ 1, & \text{if } x > 0, \end{cases}$$

see Example 1. Notice that the nonlinearities in this example are discontinuous at $x = 1$ for each $y \in \mathbb{R}_+$ and at $y = 1$ for every $x \in \mathbb{R}_+$. Moreover, the first nonlinearity has a superlinear behavior and the second one has a sublinear one. Our second approach allows to study functions which are discontinuous over time-dependent curves in $\mathbb{R}^2_+$ and the conditions imposed to these curves are local, see Definition 2 and Theorem 4. In particular, we establish the existence of a positive solution for the system

$$\begin{cases} -x''(t) = (xy)^{1/3}, \\ -y''(t) = \left(1 + (xy)^{1/3}\right) H(x^2 + y^2), \end{cases}$$

subject to the Sturm–Liouville boundary conditions.

As mentioned above, our results rely on fixed point theory for multivalued operators in cones. We finish this introductory part by recalling the version of Krasnosel'skiĭ's fixed point theorem for set-valued maps given by Fitzpatrick–Petryshyn [14].

Theorem 1. *Let X be a Fréchet space with a cone $K \subset X$. Let d be a metric on X and let $r_1, r_2 \in (0, \infty)$, $r = \min\{r_1, r_2\}$, $R = \max\{r_1, r_2\}$ and $F : \overline{B}_R(0) \cap K \longrightarrow 2^K$ usc and condensing. Suppose there exists a continuous seminorm p such that $(I - F)\left(\overline{B}_{r_1}(0) \cap K\right)$ is p-bounded. Moreover, suppose that F satisfies:*

1. *There is some $w \in K$ with $p(w) \neq 0$ and such that $x \notin F(x) + tw$ for any $t > 0$ and $x \in \partial_K B_{r_1}(0)$;*
2. *$\lambda x \notin F(x)$ for any $\lambda > 1$ and $x \in \partial_K B_{r_2}(0)$.*

Then F has a fixed point x_0 with $r \le d(x_0, 0) \le R$.

In the case of a Banach space $(X, \|\cdot\|_X)$ and of an operator $F = (F_1, F_2) : K \subset X^2 \to 2^K$ under the hypotheses of the previous theorem, we obtain the existence of a fixed point $x = (x_1, x_2)$ for F such that $r \le \|x\| \le R$, where $\|\cdot\|$ denotes a norm in X^2, for example, $\|(x_1, x_2)\| = \|x_1\|_X + \|x_2\|_X$. Then $0 \le \|x_1\|_X \le R$ and $0 \le \|x_2\|_X \le R$, but it is not possible to obtain a lower bound for the norm of every component. This fact motivates the use of a vector version of Krasnosel'skiĭ's fixed point theorem. Such a version was introduced in [15] for single-valued operators. Another advantage of the vector approach is that it allows different behaviors in each component of the system.

2. Multivalued Vector Version of Krasnosel'skiĭ's Fixed Point Theorem

In the sequel, let $(X, \|\cdot\|)$ be a Banach space, $K_1, K_2 \subset X$ two cones and $K := K_1 \times K_2$ the corresponding cone of $X^2 = X \times X$. For $r, R \in \mathbb{R}^2_+$, $r = (r_1, r_2)$, $R = (R_1, R_2)$, we denote

$$(K_i)_{r_i,R_i} := \{u \in K_i : r_i \le \|u\| \le R_i\} \quad (i = 1, 2),$$
$$K_{r,R} := \{u \in K : r_i \le \|u_i\| \le R_i \text{ for } i = 1, 2\}.$$

The following fixed point theorem is an extension of the vector version of Krasnosel'skiĭ's fixed point theorem given in [15,16] to the class of upper semicontinuous (usc, for short) multivalued mappings.

Theorem 2. *Let $\alpha_i, \beta_i > 0$ with $\alpha_i \ne \beta_i$, $r_i = \min\{\alpha_i, \beta_i\}$ and $R_i = \max\{\alpha_i, \beta_i\}$ for $i = 1, 2$. Assume that $N : K_{r,R} \to 2^K$, $N = (N_1, N_2)$, is an usc map with nonempty closed and convex values such that $N(K_{r,R})$ is compact, and there exist $h_i \in K_i \setminus \{0\}$, $i = 1, 2$, such that for each $i \in \{1, 2\}$ the following conditions are satisfied:*

$$\lambda u_i \notin N_i u \quad \text{for any } u \in K_{r,R} \text{ with } \|u_i\| = \alpha_i \text{ and any } \lambda > 1; \tag{1}$$
$$u_i \notin N_i u + \mu h_i \quad \text{for any } u \in K_{r,R} \text{ with } \|u_i\| = \beta_i \text{ and any } \mu > 0. \tag{2}$$

Then N has a fixed point $u = (u_1, u_2)$ in K, that is, $u \in Nu$, with $r_i \le \|u_i\| \le R_i$ for $i = 1, 2$.

Proof. We shall consider the four possible combinations of compression-expansion conditions for N_1 and N_2.

1. Assume first that $\beta_i < \alpha_i$ for both $i = 1, 2$ (compression for N_1 and N_2). Then $r_i = \beta_i$ and $R_i = \alpha_i$ for $i = 1, 2$. Denote $h = (h_1, h_2)$ and define the map $\tilde{N} : K \to K$ given, for $u \in K$, by

$$\tilde{N}u = \min\left\{\frac{\|u_1\|}{r_1}, \frac{\|u_2\|}{r_2}, 1\right\} N\left(\delta_1(u_1)\frac{u_1}{\|u_1\|}, \delta_2(u_2)\frac{u_2}{\|u_2\|}\right) + \left(1 - \min\left\{\frac{\|u_1\|}{r_1}, \frac{\|u_2\|}{r_2}, 1\right\}\right) h,$$

where $\delta_i(u_i) = \max\{\min\{u_i, R_i\}, r_i\}$ for $i = 1, 2$.

The map $\tilde{N}$ is usc (the composition of usc maps is usc, see [17], Theorem 17.23) and $\tilde{N}(K)$ is relatively compact since its values belong to the compact set $\overline{\text{co}}\,(N(K_{r,R}) \cup \{h\})$. Then Kakutani's fixed point theorem implies that there exists $u \in K$ such that $u \in \tilde{N}u$.

It remains to prove that $u \in K_{r,R}$. It is clear that $\|u_i\| > 0$ since $h_i \ne 0$ for $i = 1, 2$. Assume $0 < \|u_1\| < r_1$ and $0 < \|u_2\| < r_2$. If $\min\left\{\frac{\|u_1\|}{r_1}, \frac{\|u_2\|}{r_2}\right\} = \frac{\|u_1\|}{r_1}$, then

$$u \in \frac{\|u_1\|}{r_1} N\left(\frac{r_1}{\|u_1\|}u_1, \frac{r_2}{\|u_2\|}u_2\right) + \left(1 - \frac{\|u_1\|}{r_1}\right) h,$$

so

$$\frac{r_1}{\|u_1\|}u_1 \in N_1\left(\frac{r_1}{\|u_1\|}u_1, \frac{r_2}{\|u_2\|}u_2\right) + \frac{r_1}{\|u_1\|}\left(1 - \frac{\|u_1\|}{r_1}\right) h_1,$$

what contradicts (2) for $i = 1$. Analogously, we can obtain contradictions for any other point $u \notin K_{r,R}$, as done in [15,16] for single-valued maps.

2. Assume that $\beta_1 < \alpha_1$ (compression for N_1) and $\beta_2 > \alpha_2$ (expansion for N_2). Let $N_i^* : K_{r,R} \to K_i$ $(i = 1,2)$ be given by

$$N_1^* u = N_1 \left(u_1, \left(\frac{R_2}{\|u_2\|} + \frac{r_2}{\|u_2\|} - 1 \right) u_2 \right),$$
$$N_2^* u = \left(\frac{R_2}{\|u_2\|} + \frac{r_2}{\|u_2\|} - 1 \right)^{-1} N_2 \left(u_1, \left(\frac{R_2}{\|u_2\|} + \frac{r_2}{\|u_2\|} - 1 \right) u_2 \right). \tag{3}$$

   Notice that the map $N^* = (N_1^*, N_2^*)$ is in case 1, and thus N^* has a fixed point $v \in K_{r,R}$. Further, the point u defined as $u_1 = v_1$ and $u_2 = \left(\frac{R_2}{\|v_2\|} + \frac{r_2}{\|v_2\|} - 1 \right) v_2$ is a fixed point of the operator N.
3. The case $\beta_1 > \alpha_1$ (expansion for N_1) and $\beta_2 < \alpha_2$ (compression for N_2) is similar to the previous one by taking the map $N^* = (N_1^*, N_2^*)$ defined as

$$N_1^* u = \left(\frac{R_1}{\|u_1\|} + \frac{r_1}{\|u_1\|} - 1 \right)^{-1} N_1 \left(\left(\frac{R_1}{\|u_1\|} + \frac{r_1}{\|u_1\|} - 1 \right) u_1, u_2 \right), \tag{4}$$
$$N_2^* u = N_2 \left(\left(\frac{R_1}{\|u_1\|} + \frac{r_1}{\|u_1\|} - 1 \right) u_1, u_2 \right).$$

4. The case $\beta_i > \alpha_i$ for $i = 1,2$ (expansion for N_1 and N_2) reduces to case 1, if we consider the map $N^* = (N_1^*, N_2^*)$ where N_1^* is defined by (4) and N_2^*, by (3).

Therefore, the proof is over. □

Remark 1 (Multiplicity). *Although we are interested in fixed points for the operator N satisfying that both components are nonzero, if we replace conditions* (1) *and* (2) *in Theorem 2 by the following ones:*

$$\lambda u_i \notin N_i u \quad \text{for } \|u_i\| = \alpha_i,\ \|u_j\| \le R_j\ (j \neq i) \text{ and } \lambda \ge 1;$$
$$u_i \notin N_i u + \mu h_i \quad \text{for } \|u_i\| = \beta_i,\ \|u_j\| \le R_j\ (j \neq i) \text{ and } \mu \ge 0,$$

then we can achieve multiplicity results.

Indeed, if $\beta_i > \alpha_i$ for $i = 1$ or $i = 2$, then the operator N has one additional fixed point $v = (v_1, v_2)$ such that $\|v_i\| < r_i$ and $r_j < \|v_j\| < R_j$ with $j \neq i$. Furthermore, if $\beta_i > \alpha_i$ for $i = 1,2$, then N has three nontrivial fixed points. Such cases are considered in the paper [18] in connection with (p,q)-Laplacian systems.

Our purpose is to apply Theorem 2 to a multivalued regularization of a discontinuous system of single-valued operators associated to a system of differential equations with discontinuous nonlinearities. Our aim is to obtain new existence and localization results for such kind of problems.

In order to do that, we need the following definitions and results.

Let U be a relatively open subset of the cone $K := K_1 \times K_2$ and $T : \overline{U} \to K$, $T = (T_1, T_2)$, an operator not necessarily continuous. We associate to the operator T the following multivalued map $\mathbb{T} : \overline{U} \to 2^K$ given by

$$\mathbb{T} = (\mathbb{T}_1, \mathbb{T}_2), \quad \mathbb{T}_i u = \bigcap_{\varepsilon > 0} \overline{\mathrm{co}}\, T_i \left(\overline{B}_\varepsilon(u) \cap \overline{U} \right) \quad \text{for every } u \in \overline{U} \quad (i = 1,2), \tag{5}$$

where $\overline{B}_\varepsilon(u) := \{ v \in X^2 : \|u_i - v_i\| \le \varepsilon \text{ for } i = 1,2 \}$, $\overline{U}$ denotes the closure of the set U with the relative topology of K and $\overline{\mathrm{co}}$ means closed convex hull. The map $\mathbb{T}_i$ is called the closed-convex envelope of T_i and it satisfies the following properties, see [2].

Proposition 1. *Let $\mathbb{T}$ be the closed-convex envelope of an operator $T : \overline{U} \longrightarrow K$. The following properties are satisfied:*

1. *If T maps bounded sets into relatively compact sets, then $\mathbb{T}$ assumes compact values and it is usc;*

2. *If $T\,\overline{U}$ is relatively compact, then $\mathbb{T}\,\overline{U}$ is relatively compact too.*

Remark 2. *The following two statements are equivalent:*

(a) $y \in \mathbb{T}_i(u)$ $(i = 1, 2)$;

(b) *for every $\varepsilon > 0$ and every $\rho > 0$ there exist $m \in \mathbb{N}$ and a finite family of vectors $x_j \in \overline{B}_\varepsilon(u) \cap \overline{U}$ and coefficients $\lambda_j \in [0,1]$ $(j = 1, 2, \dots, m)$ such that $\sum \lambda_j = 1$ and*

$$\left\| y - \sum_{j=1}^{m} \lambda_j\, T_i x_j \right\| < \rho.$$

3. Positive Solutions of Discontinuous Systems

We study the existence and localization of positive solutions for the following second-order coupled differential system

$$\begin{cases} u_1''(t) + g_1(t) f_1(t, u_1(t), u_2(t)) = 0, \\ u_2''(t) + g_2(t) f_2(t, u_1(t), u_2(t)) = 0, \end{cases} \tag{6}$$

for $t \in I = [0,1]$, with the following boundary conditions

$$a_i u_i(0) - b_i u_i'(0) = 0, \quad c_i u_i(1) + d_i u_i'(1) = 0, \tag{7}$$

for $i = 1, 2$, where $a_i, b_i, c_i, d_i \in \mathbb{R}_+ \equiv [0, \infty)$ and $\rho_i := b_i c_i + a_i c_i + a_i d_i > 0$ for $i = 1, 2$. Assume that, for $i = 1, 2$,

(H_1) $g_i \in L^1(I)$, $g_i(t) \geq 0$ for a.e. $t \in I$ and $\int_{1/4}^{3/4} g(s)\, ds > 0$;

(H_2) $f_i : I \times \mathbb{R}_+^2 \to \mathbb{R}_+$ satisfies that

 (i) $f_i(\cdot, u_1(\cdot), u_2(\cdot))$ are measurable whenever $(u_1, u_2) \in \mathcal{C}(I)^2$;

 (ii) for each $\rho > 0$ there exists $R_{i,\rho} > 0$ such that

$$f_i(t, u_1, u_2) \leq R_{i,r} \quad \text{for } u_1, u_2 \in [0, \rho] \text{ and a.e. } t \in I.$$

Notice that condition (H_2) (i) is satisfied if $f_i(\cdot, u_1, u_2)$ is measurable for all constants u_1, u_2, and if $f_i(t, \cdot, \cdot)$ is continuous for a.a. t, which is not necessarily the case in this paper.

Let $X = \mathcal{C}(I)$ be the space of continuous functions defined on I endowed with the usual norm $\|v\| := \|v\|_\infty = \max_{t \in I} |v(t)|$ and let P be the cone of all nonnegative functions of X. A positive solution to (6)–(7) is a function $u = (u_1, u_2)$ with $u_i \in P \cap W^{2,1}(I)$, $u_i \not\equiv 0$ $(i = 1, 2)$ such that u satisfies (6) for a.a. $t \in I$ and the boundary conditions (7). The existence of positive solutions to problems (6)–(7) is equivalent to the existence of fixed points of the integral operator $T : P^2 \to P^2$, $T = (T_1, T_2)$, given by

$$(T_i u)(t) = \int_0^1 G_i(t,s) g_i(s) f_i(s, u_1(s), u_2(s))\, ds, \qquad i = 1, 2, \tag{8}$$

where $G_i(t,s)$ are the corresponding Green's functions which are explicitly given by

$$G_i(t,s) = \frac{1}{\rho_i} \begin{cases} (c_i + d_i - c_i t)(b_i + a_i s), & \text{if } 0 \leq s \leq t \leq 1, \\ (b_i + a_i t)(c_i + d_i - c_i s), & \text{if } 0 \leq t \leq s \leq 1. \end{cases}$$

Denote

$$M_i := \min\left\{ \frac{c_i + 4d_i}{4(c_i + d_i)}, \frac{a_i + 4b_i}{4(a_i + b_i)} \right\},$$

then it is possible to check the following inequalities:

$$\begin{aligned} G_i(t,s) &\leq G_i(s,s) \quad \text{for } t,s \in I, \\ M_i G_i(s,s) &\leq G_i(t,s) \quad \text{for } t \in [1/4,3/4],\ s \in I. \end{aligned}$$

Consider in X the cones K_1 and K_2 defined as

$$K_i = \{v \in P : v(t) \geq M_i \|v\|_\infty \text{ for all } t \in [1/4,3/4]\},$$

and the corresponding cone $K := K_1 \times K_2$ in X^2. Then, $T(K) \subset K$. Indeed, for $u \in K$ and $i = 1,2$,

$$\begin{aligned} M_i \|T_i u\| &= M_i \max_{t\in[0,1]} \int_0^1 G_i(t,s) g_i(s) f_i(s, u_1(s), u_2(s))\, ds \\ &\leq M_i \int_0^1 G_i(s,s) g_i(s) f_i(s, u_1(s), u_2(s))\, ds \leq \min_{t\in[1/4,3/4]} T_i u(t). \end{aligned}$$

Hence, $T_i u \in K_i$ for every $u \in K$ and $i = 1,2$.

Therefore, it must be clear that we intend to apply Theorem 2 in a subset of K to the multivalued operator $\mathbb{T}$ associated to the discontinuous operator T. Later, we shall provide conditions about the functions f_i ($i = 1,2$) which guarantee that $\text{Fix}(\mathbb{T}) \subset \text{Fix}(T)$, where $\text{Fix}(S)$ stands for the set of fixed points of the mapping S. As a consequence, we obtain some results concerning the existence of positive solutions for system (6)–(7).

Let us introduce some notations. For $\alpha_i, \beta_i > 0$ with $\alpha_i \neq \beta_i$ and $\varepsilon > 0$, we let $r_i = \min\{\alpha_i, \beta_i\}$, $R_i = \max\{\alpha_i, \beta_i\}$ ($i = 1,2$) and

$$\begin{aligned} f_1^{\beta,\varepsilon} &:= \inf\{f_1(t,u_1,u_2) : t \in [1/4,3/4],\ M_1(\beta_1 - \varepsilon) \leq u_1 \leq \beta_1 + \varepsilon,\ M_2 r_2 \leq u_2 \leq R_2\}, \\ f_2^{\beta,\varepsilon} &:= \inf\{f_2(t,u_1,u_2) : t \in [1/4,3/4],\ M_1 r_1 \leq u_1 \leq R_1,\ M_2(\beta_2 - \varepsilon) \leq u_2 \leq \beta_2 + \varepsilon\}, \\ f_1^{\alpha,\varepsilon} &:= \sup\{f_1(t,u_1,u_2) : t \in [0,1],\ 0 \leq u_1 \leq \alpha_1 + \varepsilon,\ 0 \leq u_2 \leq R_2\}, \\ f_2^{\alpha,\varepsilon} &:= \sup\{f_2(t,u_1,u_2) : t \in [0,1],\ 0 \leq u_1 \leq R_1,\ 0 \leq u_2 \leq \alpha_2 + \varepsilon\}. \end{aligned}$$

Also, denote

$$A_i := \inf_{t\in[1/4,3/4]} \int_{1/4}^{3/4} G_i(t,s) g_i(s)\, ds, \qquad B_i := \sup_{t\in[0,1]} \int_0^1 G_i(t,s) g_i(s)\, ds$$

for $i = 1,2$.

Lemma 1. *Assume that there exist $\alpha_i, \beta_i > 0$ with $\alpha_i \neq \beta_i$, $i = 1,2$, and $\varepsilon > 0$ such that*

$$B_i f_i^{\alpha,\varepsilon} < \alpha_i, \quad A_i f_i^{\beta,\varepsilon} > \beta_i \quad \text{for } i = 1,2. \tag{9}$$

Then, for each $i \in \{1,2\}$, the following conditions are satisfied:

$$\lambda u_i \notin \mathbb{T}_i u \quad \text{for any } u \in K_{r,R} \text{ with } \|u_i\|_\infty = \alpha_i \text{ and any } \lambda > 1; \tag{10}$$

$$u_i \notin \mathbb{T}_i u + \mu h_i \quad \text{for any } u \in K_{r,R} \text{ with } \|u_i\|_\infty = \beta_i \text{ and any } \mu > 0, \tag{11}$$

where h_1 and h_2 are constant functions equal to 1.

Moreover, the map $\mathbb{T}$ defined as in (5) has at least one fixed point in $K_{r,R}$.

Proof. First, observe that if $v \in K_{r,R}$, then

$$M_i r_i \leq v_i(t) \leq R_i \quad \text{for all } t \in \left[\frac{1}{4}, \frac{3}{4}\right] \quad (i = 1, 2),$$

and if $v \in \overline{B}_\varepsilon(u) \cap K_{r,R}$ for some $u \in K_{r,R}$, and $\|u_1\|_\infty = \alpha_1$, then $v_1(t) \leq \alpha_1 + \varepsilon$ for all $t \in [0,1]$ and

$$M_1(\alpha_1 - \varepsilon) \leq v_1(t) \leq \alpha_1 + \varepsilon \quad \text{for all } t \in \left[\frac{1}{4}, \frac{3}{4}\right].$$

Now we prove (10) for $i = 1$. Assume that $\|u_1\|_\infty = \alpha_1$ and let us see that $\lambda u_1 \notin \mathbb{T}_1 u$ for $\lambda > 1$. First, we shall show that given a family of vectors $v_k \in \overline{B}_\varepsilon(u) \cap K_{r,R}$ and numbers $\lambda_k \in [0,1]$ such that $\sum \lambda_k = 1$ $(k = 1, \ldots, m)$, then

$$\lambda u_1 \neq \sum_{k=1}^{m} \lambda_k T_1 v_k,$$

what implies that $\lambda u_1 \notin \operatorname{co}\left(T_1\left(\overline{B}_\varepsilon(u) \cap K_{r,R}\right)\right)$. Indeed, if not, taking the supremum for $t \in [0,1]$,

$$\begin{aligned}
\lambda \alpha_1 &\leq \sup_{t\in[0,1]} \sum_{k=1}^{m} \lambda_k \int_0^1 G_1(t,s) g_1(s) f_1(s, v_{k,1}(s), v_{k,2}(s))\, ds \\
&\leq \sum_{k=1}^{m} \lambda_k \sup_{t\in[0,1]} \int_0^1 G_1(t,s) g_1(s) f_1(s, v_{k,1}(s), v_{k,2}(s))\, ds \\
&\leq \sum_{k=1}^{m} \lambda_k f_1^{\alpha,\varepsilon} B_1 = f_1^{\alpha,\varepsilon} B_1 < \alpha_1,
\end{aligned}$$

a contradiction. Notice that if $\lambda u_1 \in \overline{\operatorname{co}}\left(T_1\left(\overline{B}_\varepsilon(u) \cap K_{r,R}\right)\right)$, then it is the limit of a sequence of functions satisfying the previous inequality and thus, as a limit, it satisfies $\lambda\, \alpha_1 \leq \alpha_1$ which is also a contradiction since $\lambda > 1$. Therefore, $\lambda u_1 \notin \mathbb{T}_1 u$ for $\lambda > 1$.

In order to prove (11) for $i = 1$, assume that $\|u_1\|_\infty = \beta_1$ and $u_1 = \sum_{k=1}^{m} \lambda_k T_1 v_k + \mu$ for some family of vectors $v_k \in \overline{B}_\varepsilon(u) \cap K_{r,R}$ and numbers $\lambda_k \in [0,1]$ such that $\sum \lambda_k = 1$ $(k = 1, \ldots, m)$ and some $\mu > 0$. Then for $t \in [1/4, 3/4]$, we have

$$\begin{aligned}
u_1(t) &= \sum_{k=1}^{m} \lambda_k \int_0^1 G_1(t,s) g_1(s) f_1(s, v_{k,1}(s), v_{k,2}(s))\, ds + \mu \\
&\geq \sum_{k=1}^{m} \lambda_k \int_{1/4}^{3/4} G_1(t,s) g_1(s) f_1(s, v_{k,1}(s), v_{k,2}(s))\, ds + \mu \\
&\geq \sum_{k=1}^{m} \lambda_k f_1^{\beta,\varepsilon} \int_{1/4}^{3/4} G_1(t,s) g_1(s)\, ds + \mu \\
&\geq f_1^{\beta,\varepsilon} A_1 + \mu > \beta_1 + \mu,
\end{aligned}$$

so $\beta_1 > \beta_1 + \mu$, a contradiction. Hence, $u_1 \notin \operatorname{co}\left(T_1\left(\overline{B}_\varepsilon(u) \cap K_{r,R}\right)\right) + \mu h_1$. As before,

$$u_1 \notin \overline{\operatorname{co}}\left(T_1\left(\overline{B}_\varepsilon(u) \cap K_{r,R}\right)\right) + \mu h_1$$

because in that case we arrive to the inequality $\beta_1 \geq \beta_1 + \mu$ for $\mu > 0$. Therefore, $u_1 \notin \mathbb{T}_1(u) + \mu h_1$.

Similarly, it is possible to prove conditions (10) and (11) for $i = 2$.

To finish, the conclusion is obtained by applying Theorem 2 to the operator $\mathbb{T}$. □

Remark 3 (Asymptotic conditions)**.** *The existence of $\alpha_i, \beta_i > 0$ with $\alpha_i \neq \beta_i$, $i = 1, 2$, and $\varepsilon > 0$ satisfying (9) is guaranteed, in the autonomous case, by the following sufficient conditions:*

(a) *f_1 has a superlinear behavior and f_2, a sublinear one, that is,*

$$\lim_{x\to\infty}\frac{f_1(x,y)}{x}=+\infty \quad \text{for all } y>0, \quad \lim_{x\to 0}\frac{f_1(x,y)}{x}=0 \quad \text{for all } y\geq 0;$$
$$\lim_{y\to\infty}\frac{f_2(x,y)}{y}=0 \quad \text{for all } x\geq 0, \quad \lim_{y\to 0}\frac{f_2(x,y)}{y}=+\infty \quad \text{for all } x>0.$$

(b) *Both f_1 and f_2 have a superlinear behavior, that is,*

$$\lim_{x\to\infty}\frac{f_1(x,y)}{x}=+\infty \quad \text{for all } y>0, \quad \lim_{x\to 0}\frac{f_1(x,y)}{x}=0 \quad \text{for all } y\geq 0;$$
$$\lim_{y\to\infty}\frac{f_2(x,y)}{y}=+\infty \quad \text{for all } x>0, \quad \lim_{y\to 0}\frac{f_2(x,y)}{y}=0 \quad \text{for all } x\geq 0.$$

(c) *Both f_1 and f_2 have a sublinear behavior, that is,*

$$\lim_{x\to\infty}\frac{f_1(x,y)}{x}=0 \quad \text{for all } y\geq 0, \quad \lim_{x\to 0}\frac{f_1(x,y)}{x}=+\infty \quad \text{for all } y>0;$$
$$\lim_{y\to\infty}\frac{f_2(x,y)}{y}=0 \quad \text{for all } x\geq 0, \quad \lim_{y\to 0}\frac{f_2(x,y)}{y}=+\infty \quad \text{for all } x>0.$$

Remark 4. *If f_1 and f_2 are monotone in both variables, it is possible to specify the numbers $f_i^{\alpha,\varepsilon}$ and $f_i^{\beta,\varepsilon}$ ($i=1,2$), so in this case, conditions (9) only depend on the behavior of the functions at four points in $\mathbb{R}^2_+$, see [15,16].*

Note that Lemma 1 gives us sufficient conditions for the existence of a fixed point in $K_{r,R}$ of the multivalued operator $\mathbb{T}$. Hence, it remains to provide hypothesis on the functions f_i ($i=1,2$) which imply $\text{Fix}(\mathbb{T})\subset\text{Fix}(T)$ in order to obtain a solution for the system (6)–(7). Observe also that no continuity hypotheses were required to the functions f_i until now.

The following definition introduces some curves where we allow the functions f_i to be discontinuous in each variable. The idea of using such curves can be found in some recent papers for second-order discontinuous scalar problems [1–3] and, in some sense, it recalls the notion of time-depending discontinuity sets from [9].

Definition 1. *We say that $\Gamma_1:[a_1,b_1]\subset I=[0,1]\to\mathbb{R}_+$, $\Gamma_1\in W^{2,1}(a_1,b_1)$, is an inviable discontinuity curve with respect to the first variable u_1 if there exist $\varepsilon>0$ and $\psi_1\in L^1(a_1,b_1)$, $\psi_1(t)>0$ for a.e. $t\in[a_1,b_1]$ such that either*

$$\Gamma_1''(t)+\psi_1(t)<-g_1(t)f_1(t,y,z) \text{ for a.e. } t\in[a_1,b_1], \text{ all } y\in[\Gamma_1(t)-\varepsilon,\Gamma_1(t)+\varepsilon] \text{ and all } z\in\mathbb{R}_+, \quad (12)$$

or

$$\Gamma_1''(t)-\psi_1(t)>-g_1(t)f_1(t,y,z) \text{ for a.e. } t\in[a_1,b_1], \text{ all } y\in[\Gamma_1(t)-\varepsilon,\Gamma_1(t)+\varepsilon] \text{ and all } z\in\mathbb{R}_+. \quad (13)$$

Similarly, we say that $\Gamma_2:[a_2,b_2]\subset I=[0,1]\to\mathbb{R}_+$, $\Gamma_2\in W^{2,1}(a_2,b_2)$, is an inviable discontinuity curve with respect to the second variable u_2 if there exist $\varepsilon>0$ and $\psi_2\in L^1(a_2,b_2)$, $\psi_2(t)>0$ for a.e. $t\in[a_2,b_2]$ such that either

$$\Gamma_2''(t)+\psi_2(t)<-g_2(t)f_2(t,y,z) \text{ for a.e. } t\in[a_2,b_2], \text{ all } y\in\mathbb{R}_+ \text{ and all } z\in[\Gamma_2(t)-\varepsilon,\Gamma_2(t)+\varepsilon],$$

or

$$\Gamma_2''(t)-\psi_2(t)>-g_2(t)f_2(t,y,z) \text{ for a.e. } t\in[a_2,b_2], \text{ all } y\in\mathbb{R}_+ \text{ and all } z\in[\Gamma_2(t)-\varepsilon,\Gamma_2(t)+\varepsilon].$$

Now we state some technical results that we need in the proof of the condition $\mathrm{Fix}(\mathbb{T}) \subset \mathrm{Fix}(T)$. Their proofs can be found in [3]. In the sequel, m denotes the Lebesgue measure in $\mathbb{R}$.

Lemma 2 ([3], Lemma 4.1). *Let $a, b \in \mathbb{R}$, $a < b$, and let $g, h \in L^1(a,b)$, $g \geq 0$ a.e., and $h > 0$ a.e. in (a,b). For every measurable set $J \subset (a,b)$ with $m(J) > 0$ there is a measurable set $J_0 \subset J$ with $m(J \setminus J_0) = 0$ such that for every $\tau_0 \in J_0$ we have*

$$\lim_{t \to \tau_0^+} \frac{\int_{[\tau_0,t]\setminus J} g(s)\,ds}{\int_{\tau_0}^t h(s)\,ds} = 0 = \lim_{t \to \tau_0^-} \frac{\int_{[t,\tau_0]\setminus J} g(s)\,ds}{\int_t^{\tau_0} h(s)\,ds}.$$

Corollary 1 ([3], Corollary 4.2). *Let $a, b \in \mathbb{R}$, $a < b$, and let $h \in L^1(a,b)$ be such that $h > 0$ a.e. in (a,b). For every measurable set $J \subset (a,b)$ with $m(J) > 0$ there is a measurable set $J_0 \subset J$ with $m(J \setminus J_0) = 0$ such that for all $\tau_0 \in J_0$ we have*

$$\lim_{t \to \tau_0^+} \frac{\int_{[\tau_0,t]\cap J} h(s)\,ds}{\int_{\tau_0}^t h(s)\,ds} = 1 = \lim_{t \to \tau_0^-} \frac{\int_{[t,\tau_0]\cap J} h(s)\,ds}{\int_t^{\tau_0} h(s)\,ds}.$$

We shall also need the following lemma, see [2], Lemma 3.11.

Lemma 3. *If $M \in L^1(0,1)$, $M \geq 0$ almost everywhere, then the set*

$$Q = \left\{ u \in C^1([0,1]) : |u'(t) - u'(s)| \leq \int_s^t M(r)\,dr \quad \text{whenever } 0 \leq s \leq t \leq 1 \right\}$$

is closed in $C([0,1])$ endowed with the maximum norm topology.

Moreover, if $u_n \in Q$ for all $n \in \mathbb{N}$ and $u_n \to u$ uniformly in $[0,1]$, then there exists a subsequence $\{u_{n_k}\}$ which tends to u in the C^1 norm.

Now we are ready to present the following existence and localization result for the differential system (6)–(7).

Theorem 3. *Suppose that the functions f_i and g_i ($i = 1,2$) satisfy conditions (H_1), (H_2) and*

(H_3) *There exist inviable discontinuity curves $\Gamma_{1,n} : I_{1,n} := [a_{1,n}, b_{1,n}] \subset I \to \mathbb{R}_+$ with respect to the first variable, $n \in \mathbb{N}$, and inviable discontinuity curves $\Gamma_{2,n} : I_{2,n} := [a_{2,n}, b_{2,n}] \subset I \to \mathbb{R}_+$ with respect to the second variable, $n \in \mathbb{N}$, such that for each $i \in \{1,2\}$ and for a.e. $t \in I$ the function $(u_1, u_2) \mapsto f_i(t, u_1, u_2)$ is continuous on*

$$\left(\mathbb{R}_+ \setminus \bigcup_{\{n : t \in I_{1,n}\}} \{\Gamma_{1,n}(t)\} \right) \times \left(\mathbb{R}_+ \setminus \bigcup_{\{n : t \in I_{2,n}\}} \{\Gamma_{2,n}(t)\} \right).$$

Moreover, assume that there exist $\alpha_i, \beta_i > 0$ with $\alpha_i \neq \beta_i$, $i = 1,2$, and $\varepsilon > 0$ such that

$$B_i f_i^{\alpha,\varepsilon} < \alpha_i, \quad A_i f_i^{\beta,\varepsilon} > \beta_i \quad \text{for } i = 1,2.$$

Then system (6)–(7) has at least one solution in $K_{r,R}$.

Proof. The operator $T : K_{r,R} \to K$, $T = (T_1, T_2)$, given by (8) is well-defined and the hypotheses (H_1) and (H_2) imply that $T(K_{r,R})$ is relatively compact as an immediate consequence of the

Ascoli–Arzelá theorem. Moreover, by (H_1) and (H_2), there exist functions $\eta_i \in L^1(I)$ $(i = 1,2)$ such that

$$g_i(t)f_i(t,u_1,u_2) \le \eta_i(t) \quad \text{for a.e. } t \in I \text{ and all } u_1 \in [0,R_1],\ u_2 \in [0,R_2]. \tag{14}$$

Therefore, $T(K_{r,R}) \subset Q_1 \times Q_2$, where

$$Q_i = \left\{ u \in \mathcal{C}^1([0,1]) : |u'(t) - u'(s)| \le \int_s^t \eta_i(r)\,dr \quad \text{whenever } 0 \le s \le t \le 1 \right\},$$

for $i = 1,2$, which by virtue of Lemma 3 is a closed and convex subset of $X = \mathcal{C}(I)$. Then, by 'convexification', $\mathbb{T}(K_{r,R}) \subset Q_1 \times Q_2$, where $\mathbb{T}$ is the multivalued map associated to T defined as in (5).

By Lemma 1, the multivalued map $\mathbb{T}$ has a fixed point in $K_{r,R}$. Hence, if we show that all the fixed points of the operator $\mathbb{T}$ are fixed points of T, the conclusion is obtained. To do so, we fix an arbitrary function $u \in K_{r,R} \cap (Q_1 \times Q_2)$ and we consider three different cases.

Case 1: $m(\{t \in I_{1,n} : u_1(t) = \Gamma_{1,n}(t)\} \cup \{t \in I_{2,n} : u_2(t) = \Gamma_{2,n}(t)\}) = 0$ for all $n \in \mathbb{N}$. Let us prove that T is continuous at u, which implies that $\mathbb{T}u = \{Tu\}$, and therefore the relation $u \in \mathbb{T}u$ gives that $u = Tu$.

The assumption implies that for a.a. $t \in I$ the mappings $f_1(t,\cdot)$ and $f_2(t,\cdot)$ are continuous at $u(t) = (u_1(t), u_2(t))$. Hence if $u_k \to u$ in $K_{r,R}$ then

$$f_i(t,u_k(t)) \to f_i(t,u(t)) \quad \text{for a.a. } t \in I \text{ and for } i = 1,2,$$

which, along with (14), yield $Tu_k \to Tu$ in $\mathcal{C}(I)^2$, so T is continuous at u.

Case 2: $m(\{t \in I_{1,n} : u_1(t) = \Gamma_{1,n}(t)\}) > 0$ for some $n \in \mathbb{N}$. In this case we can prove that $u_1 \notin \mathbb{T}_1 u$, and thus $u \notin \mathbb{T}u$.

To this aim, first, we fix some notation. Let us assume that for some $n \in \mathbb{N}$ we have $m(\{t \in I_{1,n} : u_1(t) = \Gamma_{1,n}(t)\}) > 0$ and there exist $\varepsilon > 0$ and $\psi \in L^1(I_{1,n})$, $\psi(t) > 0$ for a.a. $t \in I_{1,n}$, such that (13) holds with Γ_1 replaced by $\Gamma_{1,n}$. (The proof is similar if we assume (12) instead of (13), so we omit it.)

We denote $J = \{t \in I_{1,n} : u_1(t) = \Gamma_{1,n}(t)\}$, and we deduce from Lemma 2 that there is a measurable set $J_0 \subset J$ with $m(J_0) = m(J) > 0$ such that for all $\tau_0 \in J_0$ we have

$$\lim_{t \to \tau_0^+} \frac{2\int_{[\tau_0,t]\setminus J} \eta_1(s)\,ds}{(1/4)\int_{\tau_0}^t \psi(s)\,ds} = 0 = \lim_{t \to \tau_0^-} \frac{2\int_{[t,\tau_0]\setminus J} \eta_1(s)\,ds}{(1/4)\int_t^{\tau_0} \psi(s)\,ds}. \tag{15}$$

By Corollary 1 there exists $J_1 \subset J_0$ with $m(J_0 \setminus J_1) = 0$ such that for all $\tau_0 \in J_1$ we have

$$\lim_{t \to \tau_0^+} \frac{\int_{[\tau_0,t]\cap J_0} \psi(s)\,ds}{\int_{\tau_0}^t \psi(s)\,ds} = 1 = \lim_{t \to \tau_0^-} \frac{\int_{[t,\tau_0]\cap J_0} \psi(s)\,ds}{\int_t^{\tau_0} \psi(s)\,ds}. \tag{16}$$

Let us now fix a point $\tau_0 \in J_1$. From (15) and (16) we deduce that there exist $t_- < \tilde{t}_- < \tau_0$ and $t_+ > \tilde{t}_+ > \tau_0$, $t_\pm$ sufficiently close to τ_0 so that the following inequalities are satisfied for all $t \in [\tilde{t}_+, t_+]$:

$$2\int_{[\tau_0,t]\setminus J} \eta_1(s)\,ds < \frac{1}{4}\int_{\tau_0}^t \psi(s)\,ds, \tag{17}$$

$$\int_{[\tau_0,t]\cap J} \psi(s)\,ds \ge \int_{[\tau_0,t]\cap J_0} \psi(s)\,ds > \frac{1}{2}\int_{\tau_0}^t \psi(s)\,ds, \tag{18}$$

and for all $t \in [t_-, \tilde{t}_-]$:

$$2 \int_{[t,\tau_0]\setminus J} \eta_1(s)\, ds < \frac{1}{4} \int_t^{\tau_0} \psi(s)\, ds, \tag{19}$$

$$\int_{[t,\tau_0]\cap J} \psi(s)\, ds > \frac{1}{2} \int_t^{\tau_0} \psi(s)\, ds. \tag{20}$$

Finally, we define a positive number

$$\tilde{\rho} = \min \left\{ \frac{1}{4} \int_{\tilde{t}_-}^{\tau_0} \psi(s)\, ds, \frac{1}{4} \int_{\tau_0}^{\tilde{t}_+} \psi(s)\, ds \right\}, \tag{21}$$

and we are ready to prove that $u_1 \notin \mathbb{T}_1 u$. It suffices to prove the following claim:

Claim: let $\varepsilon > 0$ be given by our assumptions over $\Gamma_{1,n}$ as Definition 1 shows, and let $\rho = \frac{\tilde{\rho}}{2} \min\{\tilde{t}_- - t_-, t_+ - \tilde{t}_+\}$, where $\tilde{\rho}$ is as in (21). For every finite family $x_j \in \overline{B}_\varepsilon(u) \cap K_{r,R}$ and $\lambda_j \in [0,1]$ $(j = 1,2,\ldots,m)$, with $\sum \lambda_j = 1$, we have $\|u_1 - \sum \lambda_j T_1 x_j\|_\infty \geq \rho$.

Let x_j and λ_j be as in the Claim and, for simplicity, denote $y = \sum \lambda_j T_1 x_j$. For a.a. $t \in J = \{t \in I_{1,n} : u_1(t) = \Gamma_{1,n}(t)\}$ we have

$$y''(t) = \sum_{j=1}^m \lambda_j (T_1 x_j)''(t) = -\sum_{j=1}^m \lambda_j g_1(t) f_1(t, x_{j,1}(t), x_{j,2}(t)). \tag{22}$$

On the other hand, for every $j \in \{1,2,\ldots,m\}$ and every $t \in J$ we have

$$|x_{j,1}(t) - \Gamma_{1,n}(t)| = |x_{j,1}(t) - u_1(t)| < \varepsilon,$$

and then the assumptions on $\Gamma_{1,n}$ ensure that for a.a. $t \in J$ we have

$$y''(t) = -\sum_{j=1}^m \lambda_j g_1(t) f_1(t, x_{j,1}(t), x_{j,2}(t)) < \sum_{j=1}^m \lambda_j (\Gamma''_{1,n}(t) - \psi(t)) = u_1''(t) - \psi(t). \tag{23}$$

Now for $t \in [t_-, \tilde{t}_-]$ we compute

$$\begin{aligned}
y'(\tau_0) - y'(t) &= \int_t^{\tau_0} y''(s)\, ds = \int_{[t,\tau_0]\cap J} y''(s)\, ds + \int_{[t,\tau_0]\setminus J} y''(s)\, ds \\
&< \int_{[t,\tau_0]\cap J} u_1''(s)\, ds - \int_{[t,\tau_0]\cap J} \psi(s)\, ds \\
&\quad + \int_{[t,\tau_0]\setminus J} \eta_1(s)\, ds \qquad \text{(by (23), (22) and (14))} \\
&= u_1'(\tau_0) - u_1'(t) - \int_{[t,\tau_0]\setminus J} u_1''(s)\, ds - \int_{[t,\tau_0]\cap J} \psi(s)\, ds + \int_{[t,\tau_0]\setminus J} \eta_1(s)\, ds \\
&\leq u_1'(\tau_0) - u_1'(t) - \int_{[t,\tau_0]\cap J} \psi(s)\, ds + 2 \int_{[t,\tau_0]\setminus J} \eta_1(s)\, ds \\
&< u_1'(\tau_0) - u_1'(t) - \frac{1}{4} \int_t^{\tau_0} \psi(s)\, ds \quad \text{(by (19) and (20))},
\end{aligned}$$

hence $y'(t) - u_1'(t) \geq \tilde{\rho}$ provided that $y'(\tau_0) \geq u_1'(\tau_0)$. Therefore, by integration we obtain

$$y(\tilde{t}_-) - u_1(\tilde{t}_-) = y(t_-) - u_1(t_-) + \int_{t_-}^{\tilde{t}_-} (y'(t) - u_1'(t))\, dt \geq y(t_-) - u_1(t_-) + \tilde{\rho}(\tilde{t}_- - t_-).$$

So, if $y(t_-) - u_1(t_-) \le -\rho$, then $\|y - u_1\|_\infty \ge \rho$. Otherwise, if $y(t_-) - u_1(t_-) > -\rho$, then we have $y(\tilde{t}_-) - u_1(\tilde{t}_-) > \rho$ and thus $\|y - u_1\|_\infty \ge \rho$, too.

Similar computations in the interval $[\tilde{t}_+, t_+]$ instead of $[t_-, \tilde{t}_-]$ show that if $y'(\tau_0) \le u_1'(\tau_0)$ then we have $u_1'(t) - y'(t) \ge \tilde{\rho}$ for all $t \in [\tilde{t}_+, t_+]$ and this also implies $\|y - u_1\| \ge \rho$. The claim is proven.

Case 3: $m(\{t \in I_{2,n} : u_2(t) = \Gamma_{2,n}(t)\}) > 0$ for some $n \in \mathbb{N}$. In this case it is possible to prove that $u_2 \notin \mathbb{T}_2 u$. The details are similar to those in Case 2, with obvious changes, so we omit them. □

Remark 5. *Observe that Definition 1 allows to study the discontinuities of the functions f_i independently in each variable u_1 and u_2, as shown in condition (H_3).*

In addition, a continuum set of discontinuity points is possible: for instance, the function f_1 may be discontinuous at the point $u_1 = 1$ for all $u_2 \in \mathbb{R}_+$ provided that the constant function $\Gamma_1 \equiv 1$ is an inviable discontinuity curve with respect to the first variable. This fact improves the ideas given in [5] for first-order autonomous systems where "only" a countable set of discontinuity points are allowed.

Remark 6. *Notice that conditions (12) and (13) are not local in the last variable. However, the condition*

$$\inf_{t \in I, x, y \in \mathbb{R}_+} f_1(t, x, y) > 0$$

implies that any constant function stands for an inviable discontinuity curve with respect to the first variable (since condition (13) holds). Moreover, any function with strictly positive second derivative is always an inviable discontinuity curve with respect to the variable u_1 without any additional condition on f_1.

Now we illustrate our existence result by some examples.

Example 1. *Consider the coupled system*

$$\begin{cases} -x''(t) = x^2 + x^2 y^2 H(a - x) H(b - y), \\ -y''(t) = \sqrt{x} + \sqrt{y} + H(x - c) H(y - d), \end{cases} \tag{24}$$

subject to the boundary conditions (7) (replacing u_1 and u_2 by x and y, respectively) where $a, b, c, d > 0$ and H denotes the Heaviside function.

The existence of numbers α_i and β_i in the conditions of (9) is guaranteed by Remark 3 (a) since $f_1(x, y) = x^2 + x^2 y^2 H(a - x) H(b - y)$ is a superlinear function and $f_2(x, y) = \sqrt{x} + \sqrt{y} + H(x - c) H(y - d)$ is a sublinear function.

On the other hand, the function $(x, y) \mapsto f_1(x, y)$ is continuous on $(\mathbb{R}_+ \setminus \{a\}) \times (\mathbb{R}_+ \setminus \{b\})$ and the constant function $\Gamma_1 \equiv a$ stands for an inviable curve with respect to the first variable. Indeed,

$$-\Gamma_1''(t) + \frac{a^2}{8} = \frac{a^2}{8} < f_1(y, z) \quad \text{for a.a. } t \in [0, 1] \text{ and for all } y \in \left[\frac{a}{2}, \frac{3a}{2}\right] \text{ and } z \in \mathbb{R}_+,$$

hence (13) holds with $\psi_1 \equiv a^2/8$.

Moreover, the constant function $\Gamma_2 \equiv b$ is an inviable curve with respect to the second variable, according to Remark 6 since

$$\inf_{x, y \in \mathbb{R}_+} f_2(x, y) > 0.$$

Similarly, the function $f_2(x, y) = \sqrt{x} + \sqrt{y} + H(x - c) H(y - d)$ satisfies the hypothesis (H_3) in Theorem 3, so the system (7)–(24) has at least one positive solution.

Example 2. *Consider the system*

$$\begin{cases} -x''(t) = x^2 + x^2 y^2 H(a + t^2 - x) H(b + mt - y), \\ -y''(t) = \sqrt{x} + \sqrt{y} + H(x - c) H(y - d), \end{cases} \tag{25}$$

subject to the boundary conditions (7), *where* $a, b, c, d > 0$ *and* $m \in \mathbb{R}$.

Now, for a.a. $t \in I$, *the function* $(x, y) \mapsto f_1(t, x, y)$, *where*

$$f_1(t, x, y) = x^2 + x^2y^2H(a + t^2 - x)H(b + mt - y),$$

is continuous on $(\mathbb{R}_+ \setminus \{a + t^2\}) \times (\mathbb{R}_+ \setminus \{b + mt\})$ *and the curve* $\Gamma_1(t) = a + t^2$ *is inviable with respect to the first variable. Indeed,* (13) *is satisfied with* $\psi_1 \equiv 1$, *since*

$$-\Gamma_1''(t) + 1 = -1 < f_1(t, y, z) \quad \text{for a.a. } t \in [0, 1] \text{ and for all } y, z \in \mathbb{R}_+.$$

On the other hand, the curve $\Gamma_2(t) = b + mt$ *is inviable with respect to the variable* y, *according to Remark* 6, *since* $\Gamma_2''(t) \equiv 0$ *and* $\inf_{x,y \in \mathbb{R}_+} f_2(x, y) > 0$.

Therefore, Theorem 3 *ensures the existence of one positive solution for problem* (7)–(25).

Nevertheless, the conditions of Definition 1 are too strong for functions f_1 which are discontinuous at a single isolated point (x_0, y_0) or, more generally, over a curve $(\gamma_1(t), \gamma_2(t))$ for $t \in \bar{I} \subset I$. This is the motivation for another definition of the notion of discontinuity curves. This notion will be a generalization of the admissible curves presented in [2] for one equation.

Definition 2. *We say that* $\gamma = (\gamma_1, \gamma_2) : [a, b] \subset I = [0, 1] \to \mathbb{R}_+^2$, $\gamma_i \in W^{2,1}(a, b)$ $(i = 1, 2)$, *is an admissible discontinuity curve for the differential equation* $u_1'' = -g_1(t)f_1(t, u_1(t), u_2(t))$ *if one of the following conditions holds:*

(a) $\gamma_1''(t) = -g_1(t)f_1(t, \gamma_1(t), \gamma_2(t))$ *for a.e.* $t \in [a, b]$ *(then we say* γ *is viable for the differential equation),*

(b) *There exist* $\varepsilon > 0$ *and* $\psi \in L^1(a, b)$, $\psi(t) > 0$ *for a.e.* $t \in [a, b]$ *such that either*

$$\gamma_1''(t) + \psi(t) < -g_1(t)f_1(t, y, z) \quad \text{for a.e. } t \in [a, b] \text{ all } y \in [\gamma_1(t) - \varepsilon, \gamma_1(t) + \varepsilon] \text{ and all } z \in [\gamma_2(t) - \varepsilon, \gamma_2(t) + \varepsilon],$$

or

$$\gamma_1''(t) - \psi(t) > -g_1(t)f_1(t, y, z) \quad \text{for a.e. } t \in [a, b] \text{ all } y \in [\gamma_1(t) - \varepsilon, \gamma_1(t) + \varepsilon] \text{ and all } z \in [\gamma_2(t) - \varepsilon, \gamma_2(t) + \varepsilon].$$

In this case we say that γ *is inviable.*

Similarly, we can define admissible discontinuity curves for $u_2'' = -g_2(t)f_2(t, u_1(t), u_2(t))$.

Theorem 4. *Suppose that the functions* f_i *and* g_i $(i = 1, 2)$ *satisfy conditions* (H_1), (H_2) *and*

(H_3^*) *There exist admissible discontinuity curves for the first differential equation* $\gamma_n : I_n := [a_n, b_n] \to \mathbb{R}_+^2$, $n \in \mathbb{N}$, *such that for a.e.* $t \in I$ *the function* $(u_1, u_2) \mapsto f_1(t, u_1, u_2)$ *is continuous on* $\mathbb{R}_+^2 \setminus \bigcup_{\{n : t \in I_n\}} \{(\gamma_{n,1}(t), \gamma_{n,2}(t))\}$;

(H_4^*) *There exist admissible discontinuity curves for the second differential equation* $\tilde{\gamma}_n : \tilde{I}_n := [\tilde{a}_n, \tilde{b}_n] \to \mathbb{R}_+^2$, $n \in \mathbb{N}$, *such that for a.e.* $t \in I$ *the function* $(u_1, u_2) \mapsto f_2(t, u_1, u_2)$ *is continuous on* $\mathbb{R}_+^2 \setminus \bigcup_{\{n : t \in \tilde{I}_n\}} \{(\tilde{\gamma}_{n,1}(t), \tilde{\gamma}_{n,2}(t))\}$.

Moreover, assume that there exist $\alpha_i, \beta_i > 0$ *with* $\alpha_i \neq \beta_i$, $i = 1, 2$, *and* $\varepsilon > 0$ *such that*

$$B_i f_i^{\alpha,\varepsilon} < \alpha_i, \quad A_i f_i^{\beta,\varepsilon} > \beta_i \quad \text{for } i = 1, 2.$$

Then the differential system (6)–(7) *has at least one solution in* $K_{r,R}$.

Proof. Notice that in virtue of Lemma 1 it is sufficient to show that $\text{Fix}(\mathbb{T}) \subset \text{Fix}(T)$. Reasoning as in the proof of Theorem 3, if we fix a function $u \in K_{r,R} \cap (Q_1 \times Q_2)$, we have to consider three different cases.

Case 1: $m(\{t \in I_n : u(t) = \gamma_n(t)\} \cup \{t \in \tilde{I}_n : u(t) = \tilde{\gamma}_n(t)\}) = 0$ for all $n \in \mathbb{N}$. Then T is continuous at u.

Case 2: $m(\{t \in I_n : u(t) = \gamma_n(t)\}) > 0$ or $m(\{t \in \tilde{I}_n : u(t) = \tilde{\gamma}_n(t)\}) > 0$ for some γ_n or $\tilde{\gamma}_n$ inviable. Then $u \notin \mathbb{T}u$. The proof follows the ideas from Case 2 in Theorem 3.

Case 3: $m(\{t \in I_n : u(t) = \gamma_n(t)\}) > 0$ or $m(\{t \in \tilde{I}_n : u(t) = \tilde{\gamma}_n(t)\}) > 0$ only for viable curves. Then the relation $u \in \mathbb{T}u$ implies $u = Tu$. In this case the idea is to show that u is a solution of the differential system. The proof is analogus to that of the equivalent case in [2], Theorem 3.12 or [3], Theorem 4.4, so we omit it here. □

Remark 7. *Notice that, in the case of a function $(u_1, u_2) \mapsto f_1(t, u_1, u_2)$ which is discontinuous at a single point (x_0, y_0), Definition 2 requires that one of the following two conditions holds:*

(i) $f_1(t, x_0, y_0) = 0$ *for a.e.* $t \in [0, 1]$*;*

(ii) *there exist* $\varepsilon > 0$ *and* $\psi \in L^1(0,1)$, $\psi(t) > 0$ *for a.e.* $t \in I$ *such that*

$$0 < \psi(t) < g_1(t) f_1(t, x, y) \text{ for a.e. } t \in I, \text{ all } x \in [x_0 - \varepsilon, x_0 + \varepsilon] \text{ and all } y \in [y_0 - \varepsilon, y_0 + \varepsilon].$$

In particular, for (ii), it suffices that there exist $\varepsilon, \delta > 0$ *such that*

$$0 < \delta < f_1(t, x, y) \text{ for a.e. } t \in I, \text{ all } x \in [x_0 - \varepsilon, x_0 + \varepsilon] \text{ and all } y \in [y_0 - \varepsilon, y_0 + \varepsilon].$$

To finish, we present two simple examples which fall outside of the applicability of Theorem 3, but which can be studied by means of Theorem 4.

Example 3. *Consider the problem*

$$\begin{cases} -x''(t) = f_1(x,y) = (xy)^{1/3} \left(2 - \cos\left(1/((x-1)^2 + (y-1)^2)\right) H\left((x-1)^2 + (y-1)^2\right)\right), \\ -y''(t) = f_2(x,y) = (xy)^{1/3}, \end{cases} \tag{26}$$

subject to the boundary conditions (7).

It is clear that f_1 and f_2 have a sublinear behavior, see Remark 3.

The function $(x,y) \mapsto f_1(x,y)$ is continuous on $\mathbb{R}^2_+ \setminus \{(1,1)\}$ and the constant function $\gamma(t) = (\gamma_1(t), \gamma_2(t)) \equiv (1,1)$ is an inviable admissible discontinuity curve for the differential equation $-x''(t) = f_1(x,y)$ since $0 < 1/\sqrt[3]{4} \le f_1(x,y)$ for all $x \in [1/2, 3/2]$ and all $y \in [1/2, 3/2]$; and $\gamma_1''(t) = 0$.

Therefore, Theorem 4 guarantees the existence of a positive solution for problem (7)–(26).

Example 4. *Consider the following system*

$$\begin{cases} -x''(t) = f_1(x,y) = (xy)^{1/3}, \\ -y''(t) = f_2(x,y) = \left(1 + (xy)^{1/3}\right) H(x^2 + y^2), \end{cases} \tag{27}$$

subject to the boundary conditions (7).

The nonlinearities of the system have again a sublinear behavior. Now, the function $(x,y) \mapsto f_2(x,y)$ is continuous on $\mathbb{R}^2_+ \setminus \{(0,0)\}$ and the constant function $\gamma(t) = (\gamma_1(t), \gamma_2(t)) \equiv (0,0)$ is a viable admissible discontinuity curve for the differential equation.

Hence, by application of Theorem 4, one obtains that the system (7)–(27) has at least one positive solution.

Author Contributions: The authors contributed equally to this work.

Funding: R. López Pouso was partially supported by Ministerio de Economía y Competitividad, Spain, and FEDER, Project MTM2016-75140-P, and Xunta de Galicia ED341D R2016/022 and GRC2015/004. Jorge Rodríguez-López was partially supported by Xunta de Galicia Scholarship ED481A-2017/178.

Conflicts of Interest: The authors declare no conflict of interest.

References

1. Figueroa, R.; Infante, G. A Schauder–type theorem for discontinuous operators with applications to second-order BVPs. *Fixed Point Theory Appl.* **2016**, *2016*, 53. [CrossRef]
2. Figueroa, R.; Pouso, R.L.; Rodríguez-López, J. A version of Krasnosel'skiĭ's compression-expansion fixed point theorem in cones for discontinuous operators with applications. *Topol. Methods Nonlinear Anal.* **2018**, *51*, 493–510. [CrossRef]
3. López Pouso, R. Schauder's fixed–point theorem: New applications and a new version for discontinuous operators. *Bound. Value Probl.* **2012**, *2012*, 92. [CrossRef]
4. Filippov, A.F. *Differential Equations with Discontinuous Righthand Sides*; Kluwer Academic: Dordrecht, The Netherlands, 1988.
5. Hu, S. Differential equations with discontinuous right-hand sides. *J. Math. Anal. Appl.* **1991**, *154*, 377–390. [CrossRef]
6. Spraker, J.S.; Biles, D. A comparison of the Carathéodory and Filippov solution sets. *J. Math. Anal. Appl.* **1996**, *198*, 571–580. [CrossRef]
7. Bereanu, C.; Jebelean, P.; Şerban, C. The Dirichlet problem for discontinuous perturbations of the mean curvature operator in Minkowski space. *Electron. J. Qual. Theory Differ. Equ.* **2015**, *35*, 1–7. [CrossRef]
8. Bonanno, G.; Jebelean, P.; Şerban, C. Three periodic solutions for discontinuous perturbations of the vector p-Laplacian operator. *Proc. R. Soc. Edinb. A* **2017**, *147*, 673–681. [CrossRef]
9. Cid, J.A.; Pouso, R.L. Ordinary differential equations and systems with time-dependent discontinuity sets. *Proc. R. Soc. Edinb. A* **2004**, *134*, 617–637. [CrossRef]
10. Bonanno, G.; Bisci, G.M. Infinitely many solutions for a boundary value problem with discontinuous nonlinearities. *Bound. Value Probl.* **2009**, *2009*, 670675. [CrossRef]
11. Bonanno, G.; Buccellato, S.M. Two point boundary value problems for the Sturm–Liouville equation with highly discontinuous nonlinearities. *Taiwan J. Math.* **2010**, *14*, 2059–2072. [CrossRef]
12. Bonanno, G.; Iannizzotto, A.; Marras, M. On ordinary differential inclusions with mixed boundary conditions. *Differ. Integral Equ.* **2017**, *30*, 273–288.
13. Heikkilä, S.; Lakshmikantham, V. *Monotone Iterative Techniques for Discontinuous Nonlinear Differential Equations*; Marcel Dekker: New York, NY, USA, 1994.
14. Fitzpatrick, P.M.; Petryshyn, W.V. Fixed point theorems and the fixed point index for multivalued mappings in cones. *J. Lond. Math. Soc.* **1975**, *2*, 75–85. [CrossRef]
15. Precup, R. A vector version of Krasnosel'skiĭ's fixed point theorem in cones and positive periodic solutions of nonlinear systems. *J. Fixed Point Theory Appl.* **2007**, *2*, 141–151. [CrossRef]
16. Precup, R. Componentwise compression-expansion conditions for systems of nonlinear operator equations and applications. In *Mathematical Models in Engineering, Biology and Medicine*; AIP Conference Proceedings; American Institute of Physics: Melville, NY, USA, 2009; Volume 1124, pp. 284–293.
17. Aliprantis, C.D.; Border, K.C. *Infinite Dimensional Analysis. A Hitchhiker's Guide*, 3rd ed.; Springer: New York, NY, USA, 2006.
18. Infante, G.; Maciejewski, M.; Precup, R. A topological approach to the existence and multiplicity of positive solutions of (p,q)-Laplacian systems. *Dyn. Partial Differ. Equ.* **2015**, *12*, 193–215. [CrossRef]

mathematics

MDPI

Article

Existence of Solutions for Nonhomogeneous Choquard Equations Involving p-Laplacian

Xiaoyan Shi [1], Yulin Zhao [1,*] and Haibo Chen [2]

1 School of Science, Hunan University of Technology, Zhuzhou 412007, China; xiaoyshi08@126.com
2 School of Mathematics and Statistics, Central South University, Changsha 410083, China; math_chb@csu.edu.cn
* Correspondence: zhaoylch@hut.edu.cn

Received: 26 August 2019; Accepted: 15 September 2019; Published: 19 September 2019

Abstract: This paper is devoted to investigating a class of nonhomogeneous Choquard equations with perturbation involving p-Laplacian. Under suitable hypotheses about the perturbation term, the existence of at least two nontrivial solutions for the given problems is obtained using Nehari manifold and minimax methods.

Keywords: p-Laplacian; choquard equation; nonhomogeneous; nehari method; minimax methods

---

1. Introduction and Main Results

In this paper we are interested in the following generalized nonlinear Choquard equation with perturbation involving p-Laplacian

$$-\Delta_p u + V(x)|u|^{p-2}u = \left(\int_{\mathfrak{R}^N} \frac{|u(y)|^q}{|x-y|^\mu} dy\right)|u|^{q-2}u + g(x), \qquad x \in \mathfrak{R}^N \tag{1}$$

where $N \geq 3, 2 \leq p < N, 0 < \mu < N, \frac{p}{2}(2-\frac{\mu}{N}) < q < \frac{p^*}{2}(2-\frac{\mu}{N}), 0 < V \in C^1(\mathfrak{R}^N, \mathfrak{R})$, $\Delta_p = div(|\nabla u|_{p-2}\nabla u)$ is the p-Laplacian operator, and $g : \mathfrak{R}^N \to \mathfrak{R}$ is perturbation. Here $p^* = Np/(N-p)$ denotes the Sobolev conjugate of p.

The homogeneous, a.e. $g(x) \equiv 0$, which means zero is a solution of problem (1). It was investigated in [1]. A special case of problem (1) is the well-known Choquard-Pekar equation

$$-\Delta u + u = \left(\frac{1}{|x|^\mu} * |u|^2\right)u, \qquad x \in \mathfrak{R}^N \tag{2}$$

which was investigated by Pekar [2] in relationship with the quantum field theory of a polaron. In particular, when u is a solution to (2), we know that $\phi(x,t) = u(x)e^{-it}$ is a solitary wave of the following Hartree equation

$$i\frac{\partial\phi}{\partial t} = -\Delta\phi - \left(\frac{1}{|x|^\mu} * |\phi|^2\right)\phi, \text{ in } \mathfrak{R}^3 \times \mathfrak{R}_+$$

which was introduced by Choquard in 1976 to describe an electron trapped in its own hole as approximation to Hartree-Fock theory of a one-component plasma; see [3,4]. This equation was also proposed by Penrose in [5] as a model of self-gravitating matter and is usually known in that context as the nonlinear Schrödinger-Newton equation. For more details, discussion about the physical aspects of the problem we refer the readers to [6–11] and the references therein.

From a mathematical point of view, the Choquard-Pekar Equation (2) and its generalizations have been widely studied. Take for instance, Lieb [4] investigated the existence and uniqueness, up to translations, of the ground state to problem (2) by using symmetric decreasing rearrangement inequalities. Later, Lions [6] proved the existence of infinitely many radially symmetric solutions to problem (2) via critical point theory. Ackermann [12] established some existence and multiplicity results for a type of periodic Choquard-Pekar equation with nonlocal superlinear part. Further interesting results on Choquard equations may be found in [13–26], the survey [27], and the references therein.

In [15], Ma and Zhao investigated the generalized stationary nonlinear Choquard equation

$$-\Delta u+u=\left(\int_{\mathfrak{R}^N}\frac{|u(y)|^q}{|x-y|^\mu}dy\right)|u|^{q-2}u,\qquad x\in\mathfrak{R}^N \tag{3}$$

where $N\geq 3, 0<\mu<N, q\geq 2$ Under the suitable conditions on μ, N, and q, which include the classical case, they showed that every positive solution to problem (3) is radially symmetric and monotone decreasing on some point. Using the same condition, Cingolani et al. [9] treated (3) with the case where both the vector and the scalar potential have some symmetries, and they established the regularity and some decay asymptotically at infinity of the ground states to problem (3). In [28], Moroz and Van Schaftingen eliminated this restriction and in the optimal range of parameters they derived the regularity, positivity, and radial symmetry of the ground states, and also gave decay asymptotically at infinity for them.

When the potential $V(x)$ is continuous and bounded below in $\mathfrak{R}^N$, Alves and Yang [13] studied the multiplicity and concentration behavior of positive solutions for quasilinear Choquard equation involving p-Laplacian:

$$-\varepsilon^p\Delta_p u+V(x)|u|^{p-2}u=\varepsilon^{\mu-N}\left(\int_{\mathfrak{R}^N}\frac{Q(y)F(u(y))}{|x-y|^\mu}dy\right)Q(x)f(u),\ x\in\mathfrak{R}^N \tag{4}$$

where $N\geq 3, 0<\mu<N$, V, and Q are two continuous real functions in $\mathfrak{R}^N$, ε is a positive parameter and $F(t)$ be the primate function of $f(t)$, and $\Delta_p=div(|\nabla u|_{p-2}\nabla u)$ is p-Laplacian operator, $1<p<N$ In [1], suppose that the potential V and the nonlinearity f satisfy suitable assumption, Sun considered the case $\varepsilon=1$ and $Q=1$, and proved the existence of solutions in the level of mountain pass for problem (4). Further, Alves et al. [29] considered a class of generalized Choquard equation with the nonlinearities involving N-functions, and they obtained the existence of solutions for the given Choquard equation involving the Δ_Φ-Laplacian operator, where $\Delta_\Phi=div(\phi(|\nabla u|)\nabla u)$ and $\Phi:\mathfrak{R}\to\mathfrak{R}$ is a N-function. Other related results about Choquard equation involving p-Laplacian can be found in [25,30–36] and the references therein.

In 2003, Küpper et al. [37] studied the existence of positive solutions and the bifurcation point for the following Choquard equation

$$-\Delta u+u=\left(\int_{\mathfrak{R}^3}\frac{|u(y)|^2}{|x-y|}dy\right)u+\lambda g(x),\qquad x\in\mathfrak{R}^3 \tag{5}$$

where $g(x)\in H^{-1}(\mathfrak{R}^3), g(x)\geq 0, g(x)\equiv 0$. They proved that there exist positive constants λ_* and λ_{**} such that problem (5) has at least two positive solutions for $\lambda\in(0,\lambda_*)$, and no positive solution for $\lambda>\lambda_{**}$ Furthermore, they showed that $\lambda_*=\lambda_{**}$ is a bifurcation point of problem (5).

Very recently, Xie et al. [23] showed the following nonhomogeneous Choquard equation

$$-\Delta u+V(x)u=\left(\int_{\mathfrak{R}^N}\frac{|u(y)|^q}{|x-y|^\mu}dy\right)|u|^{q-2}u+g(x),\qquad x\in\mathfrak{R}^N$$

had two nontrivial solutions if $2-\mu/N<q<(2N-\mu)/(N-2)$ satisfies the following compactness condition:

(A_1) $V\in C(\mathfrak{R}^N,\mathfrak{R}^+)$ is coercive, i.e., $\lim_{|x|\to+\infty}V(x)=+\infty$.

In [24], Zhang, Xu and Zhang also investigated the bound and ground states for nonhomogeneous Choquard equation under the following assumption.

(A_2) $V \in C(\mathfrak{R}^N, \mathfrak{R}^+)$, $\inf_{\mathfrak{R}^N} V > 0$, and there exists a positive constant r such that, for any $M > 0$, $meas\{x \in \mathfrak{R}^N : |x-y| \leq r,\ V(x) \leq M\} \to 0$ as $|y| \to +\infty$, where *meas* stands for the Lebesgue measure.

In [38], Shen, Gao and Yang considered a class of critical nonhomogeneous Choquard equation

$$-\Delta u = \left(\int_{\mathfrak{R}^N} \frac{|u(y)|^{q_*}}{|x-y|^{\mu}} dy \right) |u|^{q_*-2} u + \lambda u + g(x), \qquad x \in \Omega$$

where Ω is a smooth bounded domain of $\mathfrak{R}^N$, 0 in interior of Ω, $\lambda \in \mathfrak{R}$, $0 < \mu < N$, $N \geq 7$, $q_* = (2N-\mu)/(N-2)$ is the upper critical exponent. By applying variational methods, they obtain the existence of multiple solutions for the above problem when $\lambda \in (0, \lambda_1)$, where λ_1 is the first eigenvalue of $-\Delta$. Other related results about non-homogeneous Choquard equation can be found in [1,29,33,39–43] and the references therein.

Our work is motivated by the above work [23,37,41,44] where authors used the structure of associated Nehari manifold to obtain the multiplicity of solutions for the studied problems. Concerning the nonhomogeneous problem, Wang [41] dealt with the problem (1) in the case $p = 2, V \equiv 1$ and obtained the multiple solutions of problem (1). In this paper, we investigate the nonhomogeneous problem (1) in case of $2 \leq p < N$ and extend the results in the literatures [23,24,41,44]. The used approach of our paper comes from the literatures [23,24,41]. However, owe to dealing with p-Laplacian and nonlocal terms the calculation of our problem will be more complicated.

Before giving our main results, we need the following function spaces. $W^{1,p}(\mathfrak{R}^N)$ is the usual Sobolev space with norm

$$\|u\|_1^p = \int_{\mathfrak{R}^N} (|\nabla u|_p + |u|^p) dx$$

and $L^r(\mathfrak{R}^N)$, for $1 \leq r \leq \infty$ denotes the Lebesgue space with the norm

$$\|u\|_r = \left(\int_{\mathfrak{R}^N} |u|^r dx \right)^{1/r}, \text{ if } 1 \leq r < \infty$$

In what follows, we consider the following Banach space

$$E_V = \left\{ u \in W^{1,p}(\mathfrak{R}^N) : \int_{\mathfrak{R}^N} V(x)|u|^p dx < +\infty \right\}$$

endowed with the inner product and norm

$$\langle u, v \rangle = \int_{\mathfrak{R}^N} |\nabla u|^{p-2} \nabla u \nabla v dx + \int_{\mathfrak{R}^N} |u|^{p-2} uv dx, \|u\|^p = \int_{\mathfrak{R}^N} (|\nabla u|^p + V(x)|u|^p dx$$

Throughout this paper, we assume the following condition on the function V.

(A_0) $V \in C(\mathfrak{R}^N, \mathfrak{R})$, $\inf_{x \in \mathfrak{R}^N} V(x) > 0$ and there exists a constant $M > 0$ such that $meas\{x \in \mathfrak{R}^N : V(x) \leq M\} < \infty$, where *meas* is the Lebesgue measure.

Now we recall the well-known embedding results in [45] (Lemma 2.1).

Lemma 1. *The following statements hold.*

(i) There exists a continuous embedding from $W^{1,p}(\mathfrak{R}^N)$ into $L^r(\mathfrak{R}^N)$ for any $r \in [p, p^*)$.

(ii) Under the condition (A_0) on V, the embedding from E_V into $L^r(\mathfrak{R}^N)$ is compact for any $r \in [p, p^*)$.

Denote S_r be the best constant of the embedding from E_V into $L^r(\mathfrak{R}^N)$ as

$$|u|_r \leq S_r \|u\|, \qquad \forall u \in E_V$$

To obtain our result, we make the following assumption on perturbation term g:

(G). The perturbation function $g \in L^{\frac{2Nq}{2N(q-1)+\mu}}(\Re^N)$, g is nonzero, and there is a positive constant $\alpha = \alpha(N,p,q,\mu,S_{\frac{2Nq}{2N-\mu}})$, such that $\left\|g\right\|_{\frac{2Nq}{2N(q-1)+\mu}} < \alpha$.

Obviously, if $g = 0$, then we always get a solution for problem (1) that is the trivial solution. Now, the main result of this article reads as follows.

Theorem 1. *Suppose* (A_0), $g \equiv 0$, *and* (G) *hold. Then problem (1) admits two weak solutions. One of which is a local minimum solution with the ground state energy, and another is bound state solution. In additional, if* $g \geq 0$ *then the two weak solutions are nonnegative.*

This paper is organized as follows. In Section 2, we introduce the variational setting for problem (1) and give some related preliminaries. In Section 3, we study the Palais-Smale sequences and the minimization problems. Finally, we give the proof of Theorem 1 in Section 4.

2. Variational Setting and Fibering Map Analysis

This section is devoted to stating the variational setting and giving some lemmas which will be used as tools to prove our main results. The key inequality is the following classical Hardy-Littlewood-Sobolev inequality [3].

Lemma 2. *(Hardy-Littlewood-Sobolev inequality [3]). Let* $t,s > 1$, *and* $0 < \mu < N$ *with* $\mu/N + 1/s + 1/t = 2$, $f \in L^t(R^N)$ *and* $g \in L^s(R^N)$. *Then there exists a constant* $C(N,t,\mu,s)$ *independent of* f, g *such that*

$$\int_{\Re^N}\int_{\Re^N} \frac{f(x)g(y)}{|x-y|^{\mu}} dxdy \leq C(N,t,\mu,s)|f|_{L^t} \cdot |g|_{L^s}$$

By the Hardy-Littlewood-Sobolev inequality we have that

$$\int_{\Re^N}\int_{\Re^N} \frac{|u(x)|_q |u(y)|^q}{|x-y|^{\mu}} dxdy$$

is well defined if $|u|_q \in L^t(R^N)$ for some $t > 1$ satisfying

$$\frac{\mu}{N} + \frac{2}{t} = 2$$

For we will be working in the space $W^{1,p}(\Re^N)$, by Sobolev embedding theorem we obtain that $qt \in [p, p^*]$, where $p^* = Np/(N-p)$; that is

$$\frac{p}{2}(2-\frac{\mu}{N}) \leq q \leq \frac{p^*}{2}(2-\frac{\mu}{N}) = \frac{p}{2}\left(\frac{2N-\mu}{N-p}\right)$$

Define

$$q_l := \frac{p}{2}(2-\frac{\mu}{N}), \text{ and } q^u := \frac{p}{2}\left(\frac{2N-\mu}{N-p}\right)$$

Therefore, q_l and q^u are called as lower and upper critical exponents in the sense of the Hardy-Littlewood-Sobolev inequality. We constrain our discussion only when $q \in (q_l, q^u)$ We define the energy functional corresponding to problem (1) as

$$I(u) = \frac{1}{p}\int_{\Re^N} (|\nabla u|^p + V(x)|u|^p) dx - \frac{1}{2q}\int_{\Re^N}\int_{\Re^N} \frac{|u(x)|^q |u(y)|^q}{|x-y|^{\mu}} dxdy - \int_{\Re^N} g(x)udx, u \in E_V$$

By the condition (G), Hardy-Littlewood-Sobolev inequality and Sobolev inequality, we have

$$\int_{\Re^N}\int_{\Re^N}\frac{|u(x)|^q|u(y)|^q}{|x-y|^\mu}dxdy \le C(N,\mu)|u^q|^2_{\frac{2Nq}{2N-\mu}} \le C(N,\mu)S^{2q}_{\frac{2Nq}{2N-\mu}}\|u\|^{2q} \tag{6}$$

and

$$\int_{\Re^N} g(x)udx \le |g|_{\frac{2Nq}{2N(q-1)+\mu}}|u|_{\frac{2Nq}{2N-\mu}} \le |g|_{\frac{2Nq}{2N(q-1)+\mu}} S_{\frac{2Nq}{2N-\mu}}\|u\| \tag{7}$$

for any $u^q \in L^r(\Re^N), r>1, \mu \in (0,N)$ and $q_l \le q \le q^u, g \in L^{\frac{2Nq}{2N(q-1)+\mu}}(\Re^N)$. Therefore, one knows that I is well defined and $I(u) \in C^2(E_V, \Re)$ and its critical points are weak solutions of problem (1). Moreover,

$$\langle I'(u), v\rangle = \int_{\Re^N}(|\nabla u|^{p-2}\nabla u\nabla v + V(x)|u|^{p-2}uv)dx$$

$$-\int_{\Re^N}\int_{\Re^N}\frac{|u(y)|^q|u(x)|^{q-2}u(x)v(x)}{|x-y|^\mu}dxdy - \int_{\Re^N} g(x)vdx$$

for all $v \in E_V$. Thus, we will constrain our functional I on the Nehari manifold

$$\Lambda = \{u \in E_V : \langle I'(u), u\rangle = 0\}$$

Clearly, every nontrivial weak solution of problem (1) belongs to Λ. Denote $\Psi(u) = \langle I'(u), u\rangle$, so we can see that

$$\langle I'(u), u\rangle = \|u\|^p - \int_{\Re^N}\int_{\Re^N}\frac{|u(x)|^q|u(y)|^q}{|x-y|^\mu}dxdy - \int_{\Re^N} g(x)u(x)dx$$

and

$$\langle \Psi'(u), u\rangle = p\|u\|^p - 2q\int_{\Re^N}\int_{\Re^N}\frac{|u(x)|^q|u(y)|^q}{|x-y|^\mu}dxdy - \int_{\Re^N} g(x)u(x)dx$$

Notice that, if u_0 is a local minimum solution of the functional I, one has

$$\langle I'(u_0), u_0\rangle = 0, \quad \langle \Psi'(u_0)\ , u_0\rangle \ge 0$$

Thus, we can subdivide the Nehari manifold Λ into three parts as follows:

$$\Lambda^+ = \{u \in \Lambda : \langle \Psi'(u), u\rangle > 0\}$$

$$\Lambda^- = \{u \in \Lambda : \langle \Psi'(u), u\rangle < 0\}$$

$$\Lambda^0 = \{u \in \Lambda : \langle \Psi'(u), u\rangle = 0\}$$

Clearly, only Λ^0 contains the element 0. It is easy to see that $\Lambda^0 \cup \Lambda^+$ and $\Lambda^0 \cup \Lambda^-$ are closed subsets of E_V. In the due course of this paper, we will subsequently give reason to divide the set Λ into above three subsets.

For the convenience of calculations, for $u \in E_V$, we denote

$$A := A(u) = \int_{\Re^N}(|\nabla u|^{p-1}\nabla u + V(x)|u|^{p-1}u)dx = \|u\|^p$$

$$B := B(u) = \int_{\Re^N}\int_{\Re^N}\frac{|u(x)|^q|u(y)|^q}{|x-y|^\mu}dxdy$$

$$C := C(u) = \int_{\Re^N} g(x)udx$$

For $u \in E_V$, we define the fibering map $\varphi : (0, +\infty) \to \Re$ as

$$\varphi(t) := I(tu) = \frac{A}{p}t^p - \frac{B}{2q}t^{2q} - Ct, \qquad t > 0 \tag{8}$$

From (8) we have

$$\varphi'(t) = \frac{1}{t}\langle I'(tu), tu\rangle = \frac{\Psi(tu)}{t} = At^{p-1} - Bt^{2q-1} - C \tag{9}$$

which implies that $u \in \Lambda$ if and only if $\varphi'(1) = 0$. It is easy to see that $tu \in \Lambda$ with $t > 0$ if and only if $\varphi'(t) = 0$, i.e., $\Lambda = \{u \in E_V : \varphi'(t) = 0\}$. Moreover,

$$\varphi''(t) = \frac{\langle \Psi'(tu), tu\rangle - \Psi(tu)}{t^2} = (p-1)At^{p-2} - (2q-1)Bt^{2q-2} \tag{10}$$

which implies that for $u \in \Lambda$, $\langle \Psi'(tu), tu\rangle > 0$ or < 0 if and only if $\varphi''(t) > 0$ or < 0, respectively. That is to say, from the sign of $\varphi''(t)$ the stationary points of $\varphi(t)$ can be divided into three types, namely local minimum, local maximum, and turning point. Thus, $\Lambda^{\pm}$ and Λ^0 can also be written as

$$\Lambda^{\pm} = \{tu \in \Lambda : \varphi''(t) > 0 \text{ or } < 0\}, \text{and } \Lambda^0 = \{u \in \Lambda : \varphi''(t) = 0\}$$

Lemma 3. *Assume that $g \equiv 0$ and satisfies (G). Then for any $u \in E_V \backslash \{0\}$, there exists a unique $t_1 = t_1(u) > 0$ such that $t_1u \in \Lambda^-$. In particular,*

$$t_1 > \left[\frac{(p-1)A}{(2q-1)B}\right]^{1/(2q-p)} := t_0$$

and $I(t_1u) = \max_{t \geq 0} I(tu)$ for $\int_{\Re^N} gudx \leq 0$.

Moreover, if $\int_{\Re^N} gudx > 0$, then there exist unique $0 < t_2 = t_2(u) < t_3 = t_3(u)$ such that $t_2u \in \Lambda^+$. In particular, $I(t_3u) = \max_{t \geq t_2} I(tu)$, $I(t_2u) = \min\limits_{0 \leq t \leq t_3} I(tu)$.

Proof. Set $k(t) = At^{p-1} - Bt^{2q-1}$, then $\varphi'(t) = k(t) - C$ and $k'(t) = \varphi''(t)$ Obviously, $\lim_{t \to 0^+} k(t) = 0$, $\lim_{t \to +\infty} k(t) = -\infty$ and $k(t) > 0$ for $t > 0$ sufficiently small. Due to $2q > p$, if $k'(t_0) = 0$, then $t_0 = \left(\frac{(p-1)A}{(2q-1)B}\right)^{1/(2q-p)}$. Thus, we have $k'(t) > 0$ for $t \in (0, t_0)$, and $k'(t) < 0$ for $t \in (t_0, +\infty)$.

In the case $C = \int_{\Re^N} g(x)udx \leq 0$, there exists a unique t_1 with $t_1 > t_0$ such that $k(t_1) = \int_{\Re^N} gudx$ and $k'(t_1) < 0$. Therefore,

$$\langle I'(t_1u), t_1u\rangle = At_1{}^p - Bt_1{}^{2q} - Ct_1 = t_1(At_1{}^{p-1} - Bt_1{}^{2q-1} - C) = t_1(k(t_1) - C) = 0$$

This implies $t_1u \in \Lambda$. Moreover,

$$\langle \Psi'(t_1u), t_1u\rangle = Apt_1^p - 2Bqt_1^{2q} - Ct_1 = (p-1)At_1^p - (2q-1)Bt_1^{2q} = t_1^2k'(t_1) < 0$$

which implies that $t_1u \in \Lambda^-$, and $I(t_1u) = \max_{t \geq 0} I(tu)$.

In the case $C = \int_{\Re^N} g(x)udx > 0$, for any $u \in E_1$, where $E_1 = \{u \in E_V : \|u\| = 1\}$. By the assumption (G) and $\widetilde{t_0} = \widetilde{t_0}(u) = (\frac{p-1}{(2q-1)\widetilde{B}(u)})^{1/(2q-p)}$, we have

$$\begin{aligned}
\max_{t\geq 0}\varphi'(t) & \geq \varphi'(\widetilde{t_0}) = \widetilde{t}_0^{p-1} - \widetilde{B}\widetilde{t}_0^{2q-1} - \widetilde{C} \\
& = [\frac{p-1}{(2q-1)\widetilde{B}}]^{\frac{p-1}{2q-p}} - \widetilde{B}[\frac{p-1}{(2q-1)\widetilde{B}}]^{\frac{2q-1}{2q-p}} - \widetilde{C} \\
& = [\frac{p-1}{(2q-1)\widetilde{B}}]^{\frac{p-1}{2q-p}} - \widetilde{B}[\frac{p-1}{(2q-1)\widetilde{B}}]^{\frac{2q-p}{2q-p}}[\frac{p-1}{(2q-1)\widetilde{B}}]^{\frac{p-1}{2q-p}} - \widetilde{C} \\
& = [\frac{p-1}{(2q-1)\widetilde{B}}]^{\frac{p-1}{2q-p}} - \frac{p-1}{2q-1}[\frac{p-1}{(2q-1)\widetilde{B}}]^{\frac{p-1}{2q-p}} - \widetilde{C} \\
& \geq \frac{(p-1)^{\frac{p-1}{2q-p}}(2q-p)}{(2q-1)^{\frac{2q-1}{2q-p}}B_0^{\frac{p-1}{2q-p}}} - |g|_{\frac{2Nq}{2N(q-1)+\mu}} S_{\frac{2Nq}{2N-\mu}} \\
& \geq (\alpha - |g|_{\frac{2Nq}{2N(q-1)+\mu}})S_{\frac{2Nq}{2N-\mu}} > 0
\end{aligned} \tag{11}$$

where

$$B_0 = \sup_{\|u\|=1}\int_{\Re^N}\int_{\Re^N}\frac{|u(x)|^q|u(y)|^q}{|x-y|^\mu}dxdy, \text{ and} \alpha = \alpha(N,p,q,\mu,S_{\frac{2Nq}{2N-\mu}}) := \frac{(p-1)^{\frac{p-1}{2q-p}}(2q-p)}{(2q-1)^{\frac{2q-1}{2q-p}}B_0^{\frac{p-1}{2q-p}}S_{\frac{2Nq}{2N-\mu}}}$$

From (27), we have for $u \in E_1$,

$$\lim_{t\to 0^+} k(t) = 0 < \int_{\Re^N} g(x)udx \leq |g|_{\frac{2Nq}{2N(q-1)+\mu}} S_{\frac{2Nq}{2N-\mu}}\|u\| = |g|_{\frac{2Nq}{2N(q-1)+\mu}} S_{\frac{2Nq}{2N-\mu}} \leq k(\widetilde{t_0})$$

Hence, there exist unique $0 < t_2 = t_2(u) < \widetilde{t_0} < t_3 = t_3(u)$ such that

$$k(t_2) = \int_{\Re^N} g(x)udx = k(t_3) \text{ and } k'(t_3) < 0 < k'(t_2)$$

Consequently, $t_2u \in \Lambda^+$ and $t_3u \in \Lambda^-$ It is easy to see that $\frac{d}{dt}I(tu) = \varphi'(t) = 0$ for $t = t_2$ or $t = t_3$, and $\varphi''(t) > 0$ for $t \in (0,\widetilde{t_0})$ and $\varphi''(t) < 0$ for $t \in (\widetilde{t_0}, +\infty)$. Then $I(t_3u) = \max_{t\geq t_2} I(tu), I(t_2u) = \min_{0\leq t\leq t_3} I(tu)$. This proof is completed. □

Lemma 4. *For $g \equiv 0$, the condition (G) is satisfied, then $\Lambda^0 = \{0\}$.*

Proof. To prove $\Lambda^0 = \{0\}$, we need to show that for any $u \in E_V\backslash\{0\}$, the fibering map $\varphi(t)$ has no critical point that is a turning point. For any $u \in \Lambda^-$, set $\widetilde{u} = u(\|u\|)^{-1}$, then $\widetilde{u} \in E_1$. By the proof of Lemma 3, $k(t)$ has a unique global maximum point $\bar{t}_0 = (\frac{p-1}{(2q-1)B(\widetilde{u})})^{1/(2q-p)}$, and

$$k(\bar{t}_0) = \frac{(2q-p)}{2q-1}\left[\frac{p-1}{(2q-1)B(\widetilde{u})}\right]^{\frac{p-1}{2q-p}} := k_0$$

According to (8)–(10), we deduce that if $0 < C(\widetilde{u}) < k_0$, the equation $\varphi'(t) = 0$ has exactly two roots $\bar{t}_1, \bar{t}_2$ satisfying $0 < \bar{t}_1 < \bar{t}_0 < \bar{t}_2$ and if $C(\widetilde{u}) \leq 0$, $\varphi'(t) = 0$ has only one point $\bar{t}_3$ such that $\bar{t}_3 > \bar{t}_0$. Since $\varphi''(t) = k'(t)$, we have $\varphi''(\bar{t}_1) > 0$, $\varphi''(\bar{t}_2) < 0$ and $\varphi''(\bar{t}_3) < 0$. Hence, if $0 < C(\widetilde{u}) < k_0$, then $\bar{t}_1\widetilde{u} \in \Lambda^+, \bar{t}_2\widetilde{u} \in \Lambda^-$ and if $C(\widetilde{u}) \leq 0$, then $\bar{t}_3\widetilde{u} \in \Lambda^-$. This implies $\Lambda^\pm \cap \{u \in E_V : \widetilde{u} \in E_1, 0 < C(\widetilde{u}) < k_0\} \neq \varnothing$ and $\Lambda^- \cap \{u \in E_V : \widetilde{u} \in E_1, C(\widetilde{u}) \leq 0\} \neq \varnothing$. As a consequence, we infer that $\Lambda^\pm$ are nonempty. It is easy to see that for any sign of $C(\widetilde{u})$, critical point of the fibering map $\varphi(t)$ is either a point of local maximum or local minimum which implies $\Lambda^0 = \{0\}$. Therefore, it remains to show that $k_0 > C(\widetilde{u})$. By the condition (G) and Lemma 3 we have

$$k_0 - C(\widetilde{u}) = k(\bar{t}_0) - C(\widetilde{u}) = \bar{t}_0^{p-1} - B(\widetilde{u})\bar{t}_0^{2q-1} - C(\widetilde{u}) > 0$$

This completes the proof. □

Lemma 5. *Assume the condition (G) holds, then Λ^- is closed.*

Proof. Let $cl(\Lambda^-)$ denote the closure of Λ^-. Due to $cl(\Lambda^-) \subset \Lambda^- \cup \{0\}$, it is sufficient to prove that $0 \notin cl(\Lambda^-)$ or equivalently the distance $dist(u, \Lambda^-) > 0$. Set $u \in \Lambda^-$ and denote $\widetilde{u} = u(\|u\|)^{-1}$, then $\widetilde{u} \in E_1$. Under the assumption (G) and the proof of Lemma 4, one has

$$C(\widetilde{u}) < \bar{t}_0^{p-1} - B(\widetilde{u})\bar{t}_0^{2q-1} = \left[\frac{p-1}{(2q-1)B(\widetilde{u})}\right]^{\frac{p-1}{2q-p}} - B(\widetilde{u})\left[\frac{p-1}{(2q-1)B(\widetilde{u})}\right]^{\frac{2q-1}{2q-p}} = \frac{2q-p}{2q-1}\cdot\left[\frac{p-1}{(2q-1)B(\widetilde{u})}\right]^{\frac{p-1}{2q-p}} = k_0 \tag{12}$$

Moreover, we have that if $C(\widetilde{u}) \leq 0$ then $\varphi'(t) = 0$ has only one point $\bar{t}_3 > \bar{t}_0$ such that $\bar{t}_3\widetilde{u} \in \Lambda^-$. Then we have $\bar{t}_3\widetilde{u} = u$ with $\|u\| = \bar{t}_3 > \bar{t}_0$. Also, if $0 < C(\widetilde{u}) < k_0$, the equation $\varphi'(t) = 0$ has exactly two roots $\bar{t}_1, \bar{t}_2$ with $0 < \bar{t}_1 < \bar{t}_0 < \bar{t}_2$ such that $\bar{t}_1\widetilde{u} \in \Lambda^+$ and $\bar{t}_2\widetilde{u} \in \Lambda^-$. Hence, we have $\bar{t}_2\widetilde{u} = u$ and $\|u\| = \bar{t}_2 > \bar{t}_0$. In a word, for any $u \in \Lambda^-$, we get $\|u\| > \bar{t}_0$. By (7) we know that $B(\widetilde{u})$ is bounded from above. It follows from definition of $\bar{t}_0$ that

$$\bar{t}_0 = \left[\frac{p-1}{(2q-1)B(\widetilde{u})}\right]^{1/(2q-p)} \geq \left[\frac{p-1}{(2q-1)\widetilde{B}_0}\right]^{1/(2q-p)} := \tau \tag{13}$$

where

$$\widetilde{B}_0 = \sup_{\|\widetilde{u}\|=1} \int_{\Re^N}\int_{\Re^N} \frac{|\widetilde{u}(x)|^q|\widetilde{u}(y)|^q}{|x-y|^\mu}dxdy$$

which implies that $dist(u, \Lambda^-) = \inf_{u\in\Lambda^-}\|u\| \geq \tau > 0$. Hence $cl(\Lambda^-) = \Lambda^-$ and this proves the Lemma. □

Lemma 6. *Assume (A_0) and (G) hold. Then the functional $I(u)$ is coercive and bounded below on Λ Thus $I(u)$ is bounded below on Λ^+ and Λ^-.*

Proof. Let $u \in \Lambda$, from $\langle I'(u), u\rangle = 0$ and (7) we derive that

$$\begin{aligned} I(u) &= \tfrac{1}{p}\int_{\Re^N}(|\nabla u|^p + V(x)|u|^p)dx - \tfrac{1}{2q}\int_{\Re^N}\int_{\Re^N}\tfrac{|u(x)|^q|u(y)|^q}{|x-y|^\mu}dxdy - \int_{\Re^N} g(x)udx \\ &= \tfrac{1}{p}A(u) - \tfrac{1}{2q}B(u) - C(u) \\ &= (\tfrac{1}{p} - \tfrac{1}{2q})A(u) - (1 - \tfrac{1}{2q})\int_{\Re^N} g(x)udx \\ &\geq \tfrac{2q-p}{2pq}\|u\|^p - \tfrac{2q-1}{2q}|g|_{\frac{2Nq}{2N(q-1)+\mu}} S_{\frac{2Nq}{2N-\mu}}\|u\| \end{aligned} \tag{14}$$

where $S_{\frac{2Nq}{2N-\mu}}$ denotes the best constant of the embedding from E_V into $L^{\frac{2Nq}{2N-\mu}}(\Re^N)$. It is to see that I is coercive and bounded below in the manifold Λ. This completes the proof. □

3. Minimization Problems and Palais-Smale Analysis

According to Lemma 6, we can define the following two minimization problems:

$$i^- := \inf_{u\in\Lambda^-} I(u) \tag{15}$$

$$i^{+} := \inf_{u\in\Lambda^{+}} I(u) \tag{16}$$

Clearly, if the infimum of (15) and (16) are achieved, then we can show that they produce a weak solution of our problem (1).

Lemma 7. *If u is a local minimizers of I on Λ^{+} and Λ^{-} respectively, then $I'(u) = 0$.*

Proof. If u is a local minimizers of I on $\Lambda^{\pm}$, then $\nabla(I|_{N^{\pm}})(u) = 0$. Using Theorem 4.1.1 of [46] we infer that there exists Lagrangian multiplier $\lambda \in \Re$ such that

$$\langle I'(u), u\rangle = \lambda\langle \Psi(u), u\rangle$$

Since $u \in \Lambda^{\pm}$, $\langle I'(u), u\rangle = 0$ and $\langle \Psi(u), u\rangle \neq 0$. This implies $\lambda = 0$. Thus u is a nontrivial weak solution of our problem (1). □

By Lemma 6 we know that the problem of investigating solutions of problem (1) can be translated into that of studying minimizers of (15) and (16).

Lemma 8. *Assume (A_0) and (G) are satisfied. Then the functional $I(u)$ satisfies (PS)c condition with $c \in \Re$. That is, if $\{u_n\}$ is a sequence in E_V satisfying*

$$I(u_n) \to c \text{ and } I'(u_n) \to 0, \text{ as } n \to +\infty \tag{17}$$

for some $c \in \Re$, then $\{u_n\}$ possesses a convergent subsequence.

Proof. If $\{u_n\}$ be a sequence in E_V satisfies (17), then similar to Lemma 6 we get that u_n is bounded in E_V. Since E_V is reflexive Banach space, up to a subsequence, we may assume that u_n weakly converges to u in E_V. By using compact embedding of E_V in $L^r(\Re^N)$ for $r \in [p, p^*)$, u_n strongly converges to u in $L^r(\Re^N)$. Since $q \in (q_l, q^u)$ and $p < \frac{2Nq}{2N-\mu} < p^*$, it follows from Hardy-Littlewood-Sobolev inequality that

$$\int_{\Re^N}\int_{\Re^N} \frac{|u_n(y)|^q |u_n(x)|^{q-2} u_n(x)[u_n(x) - u(x)]}{|x-y|^{\mu}} dxdy \leq C(N,\mu)|u_n|^{2q-1}_{\frac{2Nq}{2N-\mu}} |u_n - u|_{\frac{2Nq}{2N-\mu}} \to 0$$

as $n \to \infty$. Then, we also get

$$\int_{\Re^N}\int_{\Re^N} \frac{|u_n(y)|^q |u_n(x)|^{q-2} u_n(x)[u_n(x) - u(x)]}{|x-y|^{\mu}} dxdy \to 0, \qquad n \to \infty$$

Thus

$$\begin{aligned} o(1) &= \langle I'(u_n) - I'(u), u_n - u\rangle \\ &= \|u_n - u\|^p - \int_{\Re^N}\int_{\Re^N} \frac{|u_n(y)|^q |u_n(x)|^{q-2} u_n(x)[u_n(x)-u(x)]}{|x-y|^{\mu}} dxdy \\ &\quad + \int_{\Re^N}\int_{\Re^N} \frac{|u(y)|^q |u(x)|^{q-2} u(x)[u_n(x)-u(x)]}{|x-y|^{\mu}} dxdy \\ &= \|u_n - u\|_p + o(1) \end{aligned}$$

which implies that $u_n \to u$ in E_V and consequently ends the proof. □

The following result is an observation regarding the minimizers of Λ^{+} and Λ^{-}.

Lemma 9. *Assume (A_0) and (G) are satisfied. Then $i^{+} < 0$ and $i^{-} > 0$.*

Proof. Let $u \in \Lambda$, by the proof of Lemma 4 we have that if $0 < C(\widetilde{u}) < k_0$ corresponding to $\widetilde{u} = u\left(\|u\|\right)^{-1} \in E_1$, then $\varphi'(t) = 0$ has exactly two roots $\bar{t}_1, \bar{t}_2$ such that $0 < \bar{t}_1 < \bar{t}_0 < \bar{t}_2$ and $\bar{t}_1\widetilde{u} \in \Lambda^+$ and $\bar{t}_2\widetilde{u} \in \Lambda^-$. Since $\varphi'(t) = t^{p-1} - B(\widetilde{u})t^{2q-1} - C(\widetilde{u})$, we get that $\lim_{t\to 0^+}\varphi'(t) = -C(\widetilde{u}) < 0$ and $\varphi''(t) > 0$ for any $t \in (0, \bar{t}_0)$. Due to $\bar{t}_1$ is point of local minimum of $\varphi(t)$ and $\bar{t}_1 > 0$, we have that $\varphi(\bar{t}_1) < \lim_{t\to 0^+}\varphi(t) = 0$ and then $i^+ \le I(\bar{t}_1\widetilde{u}) = \varphi(\bar{t}_1) < 0$. Moreover $i := \inf_{u\in\Lambda} I(u) \le \inf_{u\in\Lambda^+} I(u) = i^+ < 0$.

Now we claim that $i^- > 0$. In fact, from (7) we know that $B \le M_0\|u\|^{2q}$, where $M_0 = C(N,\mu)S^{2q}_{\frac{2Nq}{2N-\mu}}$. This implies that there is a positive constant M_1 which is independent of u such that

$$\frac{\left(\|u\|_p\right)^{2q/(2q-p)}}{B^{p/(2q-p)}} = \frac{A^{2q/(2q-p)}}{B^{p/(2q-p)}} \ge M_1 \tag{18}$$

By the given assumption and (18) we discuss $\varphi(t_0)$ corresponding to u as

$$\begin{aligned}
\varphi(t_0) &= \tfrac{1}{p}\|u\|_p t_0^p - \tfrac{1}{2q}Bt_0^{2q} - Ct_0\\
&= \tfrac{A}{p}\left[\tfrac{(p-1)A}{(2q-1)B}\right]^{p/(2q-p)} - \tfrac{B}{2q}\left[\tfrac{(p-1)A}{(2q-1)B}\right]^{2q/(2q-p)} - C\left[\tfrac{(p-1)A}{(2q-1)B}\right]^{1/(2q-p)}\\
&= \tfrac{(2q-p)(2q+p-1)}{2pq(2q-1)}\cdot\tfrac{A^{2q/(2q-p)}}{B^{p/(2q-p)}} - C\left(\tfrac{p-1}{2q-1}\right)^{1/(2q-p)}\cdot\tfrac{A^{1/(2q-p)}}{B^{1/(2q-p)}}\\
&\ge \tfrac{(2q-p)(2q+p-1)}{2pq(2q-1)}\cdot\tfrac{A^{2q/(2q-p)}}{B^{p/(2q-p)}}\\
&\ge \tfrac{(2q-p)(2q+p-1)}{2pq(2q-1)}\cdot M_1 := M_*
\end{aligned}$$

where the positive constant M_* is independent of u. Hence,

$$i^- = \inf_{u\in\Lambda\setminus\{0\}}\max\{I(u)\} \ge \inf_{u\in\Lambda\setminus\{0\}}\varphi(t_0) \ge M_* > 0.$$

This completes the proof. □

Now we study the nature of minimizing sequences for the functional $I(u)$. Using the idea of [44] to obtain a $(PS)_{i^+}$ sequence from the minimization sequence of our problem (1). The following lemma is a consequence of Lemma 4.

Lemma 10. *Assume (A_0) and (G) hold. Then for $u \in \Lambda^+$, there exists a constant $\rho > 0$ and a differentiable function $\eta^+ : B(0,\rho) \to \Re_+ := (0,+\infty)$ such that $\eta^+(0) = 1$, $\eta^+(w)(u-w) \in \Lambda^+$, and*

$$\begin{aligned}
\langle(\eta^+)'(0), w\rangle = M^*[p\textstyle\int_{\Re^N}(|\nabla u|^{p-2}\nabla u\nabla w + V(x)|u|^{p-2}uw)dx\\
-2q\textstyle\int_{\Re^N}\int_{\Re^N}\frac{|u(y)|^q|u(x)|^{q-2}u(x)w(x)}{|x-y|^\mu}dxdy - \int_{\Re^N} gwdx]
\end{aligned} \tag{19}$$

for any $w \in B(0,\rho)$, where $B(0,\rho)$ denotes the ball centered at 0 with radius ρ, and $M^* = [(p-1)\|u\|^p - (2q-1)B(u)]^{-1}$.

Proof. Fixing a function $u \in \Lambda^+$, we define a C^1 mapping $\Phi : \Re \times E_V \to \Re$ as follows

$$\Phi(t,w) = t^{p-1}\|u-w\|^p - t^{2q-1}\int_{\Re^N}\int_{\Re^N}\frac{|(u-w)(y)|^q|(u-w)(x)|^q}{|x-y|^\mu}dxdy - \int_{\Re^N} g(x)(u-w)dx$$

Notice that $\Phi(1,0) = \langle I'(u), u\rangle = 0$. Moreover

$$\Phi(t,0) = At^{p-1} - t^{2q-1}B - C = \varphi'(t)$$

where φ is the fibering map defined in (8). Since $u \in \Lambda^+$, we have $\varphi''(1) > 0$, and then so $\Phi_t(1,0) = \varphi''(1) \neq 0$. By Applying the implicit function theorem at point (1,0), we get that there is $\rho = \rho(u) > 0$ and a differentiable function $\eta^+ : B(0,\rho) \to \Re_+$ such that $\eta^+(0) = 1$, $\eta^+(w)(u-w) \in \Lambda$ for any $w \in B(0,\rho)$, and

$$\langle (\eta^+)'(0), w\rangle = -\frac{\langle \Phi_w(1,0), w\rangle}{\Phi_t(1,0)}$$

Now we only show that $\eta^+(w)(u-w) \in \Lambda^+$ for any $w \in B(0,\rho)$. In fact, from Lemma 5 it follows that $\Lambda^- \cup \Lambda^0$ is closed, then the distance $dist(u, \Lambda^- \cup \Lambda^0) > 0$. Since the function $\eta^+(w)(u-w)$ is continuous with respect to w, taking $\rho = \rho(u) > 0$ sufficiently small, such that

$$\|\eta^+(w)(u-w) - u\| < \frac{1}{4} dist(u, \Lambda^- \cup \Lambda^0), \forall w \in B(0,\rho)$$

Then $\eta^+(w)(u-w)$ does not belong to $\Lambda^- \cup \Lambda^0$. Thus $\eta^+(w)(u-w) \in \Lambda^+$. Finally, (19) can be obtained by direct differentiating $\Phi(w, \eta^+(w)) = 0$ with respect to w.

This completes the proof. □

To derive a sequence $(PS)_{i^-}$ from the minimizing sequence of our problem (1), similar to Lemma 10 we can obtain the following proposition.

Proposition 1. *If (A_0) and (G) are satisfied. Then for $u \in \Lambda^-$, there exists a constant $\rho > 0$ and a differentiable function $\eta^- : B(0,\rho) \to \Re_+$ such that $\eta^-(0) = 1$, $\eta^-(w)(u-w) \in \Lambda^-$, and*

$$\begin{aligned} \langle (\eta^-)'(0), w\rangle = M^*[p \textstyle\int_{\Re^N} (|\nabla u|^{p-2}\nabla u \nabla w + V(x)|u|^{p-2}uw)dx \\ -2q \textstyle\int_{\Re^N}\int_{\Re^N} \frac{|u(y)|^q |u(x)|^{q-2}u(x)w(x)}{|x-y|^\mu} dxdy - \int_{\Re^N} gwdx] \end{aligned}$$

for any $w \in B(0,\rho)$, and $M^* = [(p-1)\|u\|^p - (2q-1)B(u)]^{-1}$.

Lemma 11. *If (A_0) and (G) are satisfied. There exists a positive constant M such that*

$$-\frac{(2q-p)(p-1)}{2pq}\theta^{\frac{p}{p-1}} \leq i = \inf_{u\in\Lambda} I(u) \leq -\frac{(2pq-2q-p)(2q-p)}{4pq^2}\cdot M \tag{20}$$

where $\theta = \frac{2q-1}{2q-p}|g|_{\frac{2Nq}{2N(q-1)}} S_{\frac{2Nq}{2N-\mu}}$

Proof. For any $u \in \Lambda$, According to (13) one has

$$I(u) \geq \frac{2q-p}{2pq}\|u\|^p - \frac{2q-1}{2q}|g|_{\frac{2Nq}{2N(q-1)+\mu}} S_{\frac{2Nq}{2N-\mu}}\|u\| \geq -\frac{(2q-p)(p-1)}{2pq}\theta^{\frac{p}{p-1}}$$

Thus,

$$i \geq -\frac{(2q-p)(p-1)}{2pq}\theta^{\frac{p}{p-1}}$$

On the other hand, set $u_0 \in \Lambda$ be the unique solution of the following equation

$$-\Delta_p u + V(x)\big|u\big|_{p-1}u = g(x), \qquad \forall x \in \Re^N$$

Due to $g \neq 0$, $\int_{\Re^N} g(x)u_0 dx = \|u_0\|^p > 0$. Then by Lemma 4, there exists $t_1 > 0$ such that $t_1 u_0 \in \Lambda^+$. Therefore,

$$\begin{aligned} I(t_1u_0) &= \tfrac{1-p}{p}\|u_0\|^p t_1^p + \tfrac{2q-1}{2q} t_1^{2q} B(u_0) \\ &< \tfrac{1-p}{p}\|u_0\|^p t_1^p + \tfrac{p(2q-1)}{4q^2} t_1^p \|u_0\|^p \\ &= -\tfrac{(2pq-2q-p)(2q-p)}{4pq^2} t_1^p \|u_0\|^p < 0 \end{aligned}$$

Choose $M = t_1^p \left\|u_0\right\|^p$ we obtain the result. □

Lemma 12. *If (A_0) and (G) are satisfied, then there exists a sequence $\{u_n\} \subset \Lambda^+$ such that $I(u_n) \to i^+$ and $I'(u_n) \to 0$ as $n \to \infty$.*

Proof. From Lemma 6, we already show that I is bounded from below on Λ, and $\Lambda^+ \cup \{0\}$ is closed in Λ. Obviously Ekeland's variational principle (see [44]) applies to the minimization problem (16). It admits a minimizing sequence $\{u_n\} \subset \Lambda^+$ such that

(i) $I(u_n) < \inf_{u\in\Lambda^+\cup\{0\}}\{I(u)\} + \frac{1}{n}$, and

(ii) $I(w) \geq I(u_n) - \frac{1}{n}\left\|w - u_n\right\|, \forall w \in \Lambda^+ \cup \{0\}$

Then by (i) we have

$$I(u) = \frac{2q-p}{p}\|u_n\|^p - \frac{2q-1}{2q}\int_{\Re^N} g(x)u_n dx < i + \frac{1}{n} \tag{21}$$

for n large enough. This together with Lemma 11 shows

$$\int_{\Re^N} g(x)u_n dx \geq \frac{(2pq-2q-p)(2q-p)}{2pq(2q-1)}M > 0 \tag{22}$$

which implies $u_n \neq 0$ for any n. By Lemma 4, we know $i \leq \inf_{u\in\Lambda^+} I(u) = i^+ < 0$. Notice that $I(0) = 0$, then $\inf_{u\in\Lambda^+\cup\{0\}}\{I(u)\} = i^+$. Hence $I(u_n) \to i^+$ as $n \to \infty$, and we can assume that $u_n \in \Lambda^+$. Then $\|u_n\|^p = B(u_n) + C(u_n)$. Furthermore, we deduce from (13) and (i) that

$$i^+ + \frac{1}{n} \geq I(u_n) \geq \frac{2q-p}{2pq}\|u\|^p - \frac{2q-1}{2q}|g|_{\frac{2Nq}{2N(q-1)+\mu}} S_{\frac{2Nq}{2N-\mu}} \|u\| \tag{23}$$

which implies that $\{u_n\}$ is bounded. Now we claim that $\inf_n \|u_n\| \geq \xi > 0$ for some constant ξ. In fact, if not, by (23), $I(u_n) \to 0$, as $n \to \infty$. Using (23) which is a contradiction to first assertion. Therefore, there exist positive constants $\xi_2 > \xi_1$ such that

$$\xi_1 \leq \|u_n\| \leq \xi_2 \tag{24}$$

Now to finish the proof, we only need to show that $I'(u_n) \to 0$, as $n \to \infty$. By Lemma 10, for each n, we get a differentiable function $\eta_n^+ : B(0, \varepsilon) \to \Re_+$ for $\varepsilon > 0$ as follows

$$\eta_n^+(\delta) := \eta_n^+(\delta h_n), -\varepsilon < \delta < \varepsilon$$

where $h_n = \frac{I'(u_n)}{\|I'(u_n)\|}$. According to Lemma 10, we get $\eta_n^+(0) = 1$, and

$$w_\delta := \eta_n^+(\delta)[u_n - \delta h_n] \in \Lambda^+$$

By Taylor's expansion and (ii), since $w_\delta \in \Lambda^+$ we have

$$\begin{aligned} \tfrac{1}{n}\left\|w_\delta - u_n\right\| &\geq I(u_n) - I(w_\delta) \\ &= (1 - \eta_n^+(\delta))\langle I'(w_\delta), u_n\rangle + \delta\eta_n^+(\delta)\langle I'(w_\delta), h_n\rangle + o(\left\|w_\delta - u_n\right\|) \end{aligned}$$

which implies

$$(\frac{1}{n}+o(1))\|w_\delta-u_n\|\geq(1-\eta_n^+(\delta))\langle I'(w_\delta),u_n\rangle+\delta\eta_n^+(\delta)\langle I'(w_\delta),h_n\rangle \tag{25}$$

Dividing (25) by δ for $\delta\neq 0$ and passing to the limit as $\delta\to 0$, we obtain

$$(\frac{1}{n}+o(1))(1+\left|(\eta_n^+)'(0)\right|\|u_n\|)\geq-(\eta_n^+)'(0)\langle I'(u_n),u_n\rangle+\|I'(u_n)\| \tag{26}$$

Since $u_n\in\Lambda^+$, it follows from (26) that

$$\|I'(u_n)\|\leq(\frac{1}{n}+o(1))(1+\left|(\eta_n^+)'(0)\right|\cdot\|u_n\|) \tag{27}$$

From (24) we know that u_n is bounded. Then it remains to prove that $\left|(\eta_n^+)'(0)\right|$ is uniformly bounded with respect to n. In fact, according to the definition of η_n^+ and Lemma 5, we have

$$\begin{aligned}\langle(\eta_n^+)'(0),h_n\rangle=&\frac{1}{(p-1)\|u_n\|_p-(2q-1)B(u_n)}[p\int_{\Re^N}(|\nabla u_n|^{p-2}\nabla u_n\nabla h_n+V(x)|u_n|^{p-2}u_nh_n)dx\\&-2q\int_{\Re^N}\int_{\Re^N}\frac{|u_n(y)|^q|u_n(x)|^{q-2}u_n(x)h_n(x)}{|x-y|^\mu}dxdy-\int_{\Re^N}gh_ndx]\end{aligned} \tag{28}$$

By the boundedness of u_n and (28), we say that there exists a constant λ such that

$$\left|(\eta_n^+)'(0)\right|=\left|\left\langle(\eta_n^+)'(0),h_n\right\rangle\right|\leq\frac{\lambda}{(p-1)\left\|u_n\right\|_p-(2q-1)B(u_n)}$$

Therefore, it remains to show that $\chi(u_n):=(p-1)\left\|u_n\right\|_p-(2q-1)B(u_n)$ possesses a positive lower bound.

To prove the existence of positive lower bound of $\chi(u_n)$, passing to a subsequence, we assume

$$\chi(u_n)=(p-1)\|u_n\|^p-(2q-1)B(u_n)=o(1),n\to\infty \tag{29}$$

Since $u_n\in\Lambda^+$, we obtain

$$\|u_n\|^p-B(u_n)=C(u_n)$$

This along with (29) gives

$$C(u_n)=\frac{2q-p}{2q-1}\|u_n\|^p+o(1) \tag{30}$$

It follows from the condition (G) that there is a sufficiently small $\mu>0$ such that $\left|g\right|_{\frac{2Nq}{2N(q-1)+\mu}}\leq(1-\mu)\alpha$. Similarly to the proof of (12), we have

$$C(u)<\frac{2q-p}{2q-1}(1-\mu)\left(\frac{p-1}{(2q-1)B(u)}\right)^{\frac{p-1}{2q-p}} \tag{31}$$

for any $u\in E_1$. Therefore, by the principle of homogeneity,

$$\frac{2q-p}{2q-1}+\frac{o(1)}{\|u_n\|^p}=\frac{C(u_n)}{\|u_n\|^p}<\frac{2q-p}{2q-1}(1-\tau)\left(\frac{(p-1)\|u_n\|^p}{(2q-1)B(u_n)}\right)^{\frac{p-1}{2q-p}} \tag{32}$$

If $\|u_n\|\to 0$, then similar to (7) one has $C(u_n)\to 0$. Therefore

$$i^++o_n(1)=I(u_n)-\frac{1}{2q}\langle I'(u_n),u_n\rangle=\frac{2q-p}{2pq}\|u_n\|^p-\frac{2q-1}{2q}C(u_n)\to 0$$

which is a contradiction with $i^+ < 0$. Thus $\|u_n\| \to 0$, as $n \to \infty$. Consequently, from (30)–(32) we can deduce that

$$\frac{2q-p}{2q-1} \leq \frac{2q-p}{2q-1}(1-\mu), n \to \infty$$

which is a contradiction. Therefore, we conclude that $I'(u_n) \to 0$, as $n \to \infty$. The proof is completed. □

Proposition 2. *Under assumptions* (A_0) *and* (G)*, there exists a sequence* $\{\hat{u}_n\} \subset \Lambda^-$ *such that* $I(\hat{u}_n) \to i^-$ *and* $I'(\hat{u}_n) \to 0$ *as* $n \to \infty$.

Proof. By Lemma 5 we know that Λ^- is closed. Thus, by Ekeland's variational principle on Λ^- we get a sequence $\{\hat{u}_n\} \subset \Lambda^-$ such that

(iii) $I(\hat{u}_n) < \inf_{u\in\Lambda^-}\{I(u)\} + \frac{1}{n}$, and (iv) $I(w) \geq I(\hat{u}_n) - \frac{1}{n}\|w - \hat{u}_n\|, \forall w \in \Lambda^-$.

From (24) we know that $\hat{u}_n$ is bounded. By coercivity of I, $\{\hat{u}_n\}$ forms a bounded sequence in Λ. Moreover, from Lemma 5 we know that $\inf_{u\in\Lambda^-}\|u\| \geq \tau > 0$, which implies that Λ^- stays away from the origin. Then using Proposition 1 and following the proof of Lemma 12 we conclude the result. □

4. The Proof of Theorem 1

In this section, we show that the minimums are achieved for i^+ and i^-, and also give the proof of Theorem 1.

Proposition 3. *Assume* $g \neq 0, (A_0)$ *and* (G) *are satisfied. Then i can be achieved at point* $u_* \in \Lambda$*, which is a weak solution of problem (1). Moreover,* $u_* \in \Lambda^+$ *and* u_* *is a local minimum for I on* E_V.

Proof. By Lemma 8, there exists a sequence $\{u_n\} \subset \Lambda$ such that $I(u_n) \to i$ and $I'(u_n) \to 0$ as $n \to \infty$. Set u_* be the weak limit of the sequence $\{u_n\}$ on E_V, then $u_n \in \Lambda$ satisfies (22) we get

$$\int_{\Re^N} g(x)u_*(x)dx > 0 \tag{33}$$

On the other hand, $I'(u_n) \to 0$ as $n \to \infty$ implies that

$$\langle I'(u_*), v\rangle = 0, \text{ for every } v \in \Lambda$$

i.e., u_* is a weak solution of problem (1). In particular, $u_* \in \Lambda$, and

$$i \leq I(u_*) \leq \lim_{n\to+\infty} \inf\{I(u_n)\} = i$$

This implies that u_* is the minimum of I over E_V.

For $u_* \in \Lambda$ be such that $i = I(u_*)$, using Lemma 9 we have $I(u_*) < 0$. Then we get $u_* \neq 0$. Therefore u_* is a nontrivial weak solution of problem (1). Since (33) holds, applying Lemma 4 we see that there exist $t_1, t_2 > 0$ such that $u_1 := t_1u_* \in \Lambda^+$ and $t_2u_* \in \Lambda^-$. We claim that $t_1 = 1$ i.e., $u_* \in \Lambda^+$. If $t_1 < 1$, then $t_2 = 1$ which means $u_* \in \Lambda^-$. Now $I(t_1u_*) \leq I(u_*) = i < 0$ which is a contradiction with $t_1u_* \in \Lambda^+$.

Next we will prove that u_* is also a local minimum of I on E_V. Obviously, for any $u \in \Lambda$ with $C(u) > 0$ we can deduce that

$$I(\widetilde{t_2}u) \leq I(\widetilde{t}u) \text{ for any } \widetilde{t} \in (0, t_0)$$

where $t_0 = \left(\frac{(p-1)A}{(2q-1)B}\right)^{1/(2q-p)}$, $\widetilde{t_2}$ is corresponding to u. Moreover, if $u = u_*$ then

$$\widetilde{t_2} = 1 < \hat{t}_0 = \left[\frac{(p-1)A(u_*)}{(2q-1)B(u_*)}\right]^{1/(2q-p)}$$

Taking $\rho > 0$ small enough so that

$$1 < \hat{t}_w = \left[\frac{(p-1)A(u_* - w)}{(2q-1)B(u_* - w)}\right]^{1/(2q-p)} \|w\| < \rho \tag{34}$$

Thus, it follows from Lemma 10 that there exists a differentiable map $\eta^+ : B(0,\rho) \to \Re_+$ such that $\eta^+(w)(u_* - w) \in \Lambda^+$ for $\|w\| < \rho$ small. Then for any $\tilde{t} \in (0, \hat{t}_w)$ we have

$$I(\tilde{t}(u_* - w)) \geq I(\eta^+(w)(u_* - w)) \geq I(u_*) \tag{35}$$

Since (34) holds, taking $\tilde{t} = 1$ in (35) we get $I(u_*) \leq I(u_* - w)$ for $\|w\| < \rho$, which implies that u_* is a local minimum of I on E_V. The proof is completed. □

Proof of Theorem 1. Firstly, we deal with the minimization problem (16). According to Proposition 3, we only need to show that there exist a nonnegative solution on Λ^+ if $g \geq 0$. Assume $g \geq 0$, from the proof of Lemma 3, it is easy to see that $B(u_*) = B(|u_*|)$ and $C(u_*) \leq C(|u_*|)$. Moreover, it follows from the proof of Lemma 4 that there exists $t_1 > 0$ such that $t_1|u_*| \in \Lambda^+$ and $t_1|u_*| > 0$. If $\varphi_u(t)$ denotes the fibering map corresponding to $u \in E_V$ as introduced in Section 2, we have $\varphi'_{|u_*|}(1) \leq \varphi'_{u_*}(1) = 0$. Since t_1 is the point of local minimum of $\varphi_{|u_*|}(t)$ for $t \in (0, t_0(|u_*|))$, where

$$t_0(|u_*|) = \left[\frac{(p-1)A(|u_*|)}{(2q-1)B(|u_*|)}\right]^{1/(2q-p)}$$

and $t_1 \geq 1$. Consequently, we have that $I(t_1|u_*|) \leq I(|u_*|)$. Then

$$i^+ \leq I(t_1|u_*|) \leq I(|u_*|) \leq I(u_*) = i^+$$

This means that $t_1|u_*|$ solves the minimization problem (16). Therefore, we find a nonnegative solution for problem (1) using the maximum principle.

Now we show that the infimum i^- is achieved and the minimizer is second weak solution of problem (1). Consider the minimization problem (15). From Proposition 2, we know that there exists a sequence $\{\hat{u}_n\} \subset \Lambda^-$ such that $I(\hat{u}_n) \to i^-$ and $I'(\hat{u}_n) \to 0$ as $n \to \infty$. By Lemma 4, we get that there exists $\tilde{u}_* \in cl(\Lambda^-) = \Lambda^-$ such that $I(\tilde{u}_*) = i^-, I'(\tilde{u}_*) = 0$. Therefore, Lemma 7 implies that $\tilde{u}_*$ is a weak solution of problem (1). In addition, if $g \geq 0$, it follows from the proof of Lemma 4 and Proposition 1 that there exists $t_2 > 0$ such that $t_2|\tilde{u}_*| \in \Lambda^-$. Let

$$t_0(|\tilde{u}_*|) = \left[\frac{(p-1)A(|\tilde{u}_*|)}{(2q-1)B(|\tilde{u}_*|)}\right]^{1/(2q-p)}$$

then since $\tilde{u}_* \in \Lambda^-$, taking account of the graph of the fibering map corresponding to $\tilde{u}_*$ we can deduce that

$$i^- \leq I(t_2|\tilde{u}_*|) \leq I(t_2\tilde{u}_*) \leq \max_{t \geq t_0(|\tilde{u}_*|)}\{I(t_2\tilde{u}_*)\} = I(\tilde{u}_*) = i^-$$

This means that $t_2|\tilde{u}_*|$ solves the minimization problem (15) and then we know that it is a nonnegative weak solution of problem (1) using the maximum principle again. Due to $\Lambda^+ \cap \Lambda^- = \varnothing$ and Lemma 9 shows that $i^+ < i^-$, then $u_* \neq \tilde{u}_*$. This ends the proof. □

5. Conclusions

In this work, we study a class of nonhomogeneous Choquard equations with perturbation involving p-Laplacian. We give sufficient conditions of the existence of at least two nontrivial solutions for problems (1). Next it is worth investigating infinitely many solutions for nonhomogeneous Choquard equations involving p-Laplacian.

Author Contributions: Supervision, Y.Z. and H.C.; Writing-original draft, X.S.; Writing-review & editing, X.S. and Y.Z.

Funding: The research was supported by Hunan Provincial Natural Science Foundation of China (No.2019JJ40068).

Acknowledgments: The authors thank the anonymous referees for their careful reading and helpful suggestions, which help to improve the quality of this paper.

Conflicts of Interest: The authors declare no conflict of interest.

References

1. Sun, X. The existence of solutions for Choquard type equation. *Acta Math Sci.* **2018**, *38*, 54–60.
2. Pekar, S. *Untersuchung Uber Die Elektronentheorie der Kristalle*; Akademie Verlag: Berlin, Germany, 1954.
3. Lieb, E.H.; Loss, M. *Analysis. Graduate Studies in Mathematics*; American Mathematical Society: Providence, RI, USA, 2001.
4. Lieb, E.H. Existence and uniqueness of the minimizing solution of Choquard's nonlinear equation. *Stud. Appl. Math.* **1977**, *57*, 93–105. [CrossRef]
5. Penrose, R. Quantum computation, entanglement and state reduction. *Philos. Trans. Roy. Soc.* **1998**, *356*, 1927–1939. [CrossRef]
6. Lions, P.L. The Choquard equation and related questions. *Nonlinear Anal.* **1980**, *4*, 1063–1072. [CrossRef]
7. Choquard, P.; Stubbe, J.; Vuray, M. Stationary solutions of the Schröinger-Newton model-an ODE approach. *Differ. Integral Equ.* **2008**, *21*, 665–679.
8. Moroz, I.M.; Penrose, R.; Tod, P. Spherically-symmetric solutions of the Schröinger-Newton Equations. *Class. Quantum Gravity* **1998**, *15*, 2733–2742. [CrossRef]
9. Cingolani, S.; Clapp, M.; Secchi, S. Multiple solutions to a magnetic nonlinear Choquard equation. *Z. Angew. Math. Phys.* **2012**, *63*, 233–248. [CrossRef]
10. Cingolani, S.; Clapp, M.; Secchi, S. Intertwining semiclassical solutions to a Schrödinger-New ton system. *Discret. Contin. Dyn. Syst.* **2013**, *6*, 891–908.
11. Cingolani, S.; Secchi, S. Ground states for the pseudo relativistic Hartree equation with external potential. *Proc. R. Soc. Edinb. Sect. A* **2015**, *145*, 73–90. [CrossRef]
12. Ackermann, N. On a periodic Schrödinger equation with nonlocal superlinear part. *Math. Z.* **2004**, *248*, 423–443. [CrossRef]
13. Alves, C.O.; Yang, M.B. Existence of semiclassical ground state solutions for a generalized Choquard equation. *J. Differ. Equ.* **2014**, *257*, 4133–4164. [CrossRef]
14. Moroz, V.; Van Schaftingen, J. Nonexistence and optimal decay of super solutions to Choquard equations in exterior domains. *J. Differ. Equ.* **2013**, *254*, 3089–3145. [CrossRef]
15. Ma, L.; Zhao, L. Classication of positive solitary solutions of the nonlinear Choquard equation. *Arch. Ration. Mech. Anal.* **2010**, *195*, 455–467. [CrossRef]
16. Moroz, V.; Van Schaftingen, J. Existence of ground states for a class of nonlinear Choquard Equations. *T. Am. Math. Soc.* **2015**, *367*, 6557–6579. [CrossRef]
17. Gao, F.S.; Yang, M.B. The Brezis-Nirenberg type critical problem for nonlinear Choquard Equation. *Sci. China Math.* **2018**, *61*, 1219–1242. [CrossRef]
18. Clapp, M.; Salazar, D. Positive and sign changing solutions to a nonlinear Choquard equation. *J. Math. Anal. Appl.* **2013**, *407*, 1–15. [CrossRef]
19. Alves, C.O.; Yang, M.B. Investigating the multiplicity and concentration behavior of solutions for quasilinear Choquard equation via penalization method. *Proc. R. Soc. Edinb. Sect. A* **2016**, *146*, 23–58. [CrossRef]
20. Alves, C.O.; Cassani, D.; Tarsi, C.; Yang, M. Existence and concentration of ground state solutions for a critical nonlocal Schrödinger equation in R^2. *J. Differ. Equ.* **2016**, *261*, 1933–1972. [CrossRef]
21. Alves, C.O.; Gao, F.; Squassina, M.; Yang, M. Singularly perturbed critical Choquard equations. *J. Differ. Equ.* **2017**, *263*, 3943–3988. [CrossRef]
22. Su, Y.; Chen, H.B. The minimizing problem involving p-Laplacian and Hardy-Littlewood-Sobolev upper critical exponent. *arXiv* **2018**, arXiv:1805.10986. [CrossRef]

23. Xie, T.; Xiao, L.; Wang, J. Existence of multiple positive solutions for Choquard equation with perturbation. *Adv. Math. Phys.* **2015**, *2015*, 760157. [CrossRef]
24. Zhang, H.; Xu, J.X.; Zhang, F.B. Bound and Ground states for a concave-convex generaliezed Choquard equation. *Acta Appl. Math.* **2017**, *147*, 81–93. [CrossRef]
25. Wang, T. Ground state solutions for Choquard type equations with a singular potential. *Electron. J. Differ. Equ.* **2017**, *52*, 1–14.
26. Li, F.Y.; Long, L.; Huang, Y.Y.; Zhang, Z.P. Ground state for Choquard equation with doubly critical growth nonlinearity. *Electron. J. Qual. Theory Differ. Equ.* **2019**, *33*, 1–15. [CrossRef]
27. Moroz, V.; Van Schaftingen, J. A guide to the Choquard equation. *J. Fixed Point Theory Appl.* **2017**, *19*, 773–813. [CrossRef]
28. Moroz, V.; Van Schaftingen, J. Ground states of nonlinear Choquard equations: Existence, qualitative properties and decay asymptotics. *J. Funct. Anal.* **2013**, *265*, 153–184. [CrossRef]
29. Alves, C.O.; Râdulescu, V.D.; Tavares, L.S. Generalized Choquard Equations Driven by Nonhomogeneous Operators. *Mediterr. J. Math.* **2019**, *16*, 20. [CrossRef]
30. Azzollini, A.; d'Avenia, P.; Pomponio, A. Quasilinear elliptic equations in R^N via variational methods and Orlicz-Sobolev embeddings. *Calc. Var. Partial Differ. Equ.* **2014**, *49*, 197–213. [CrossRef]
31. Alves, C.O.; Da Silva, A.R. Multiplicity and concentration behavior of solutions for a quasilinear problem involving N-functions via penalization method. *Electron. J. Differ. Equ.* **2016**, *158*, 1–24.
32. Alves, C.O.; Da Silva, A.R. Multiplicity and concentration of positive solutions for a class of quasilinear problems through Orlicz-Sobolev space. *J. Math. Phys.* **2016**, *57*, 11502. [CrossRef]
33. Tuhina, M.; Konijeti, S. On doubly nonlocal p-fractional coupled elliptic system, Topol. *Meth. Nonlinear Anal.* **2018**, *51*, 609–636.
34. Abdellaoui, B.; Attar, A.; Bentifour, R. On the fractional p-Laplacian equations with weight and general datum. *Adv. Nonlinear Anal.* **2019**, *8*, 144–174. [CrossRef]
35. Pucci, P.; Xiang, M.Q.; Zhang, B.L. Existence results for Schrödinger-Choquard-Kirchhoff equations involving the fractional p-Laplacian. *Adv. Nonlinear Anal.* **2019**, *12*, 253–275. [CrossRef]
36. Ambrosio, V. On the multiplicity and concentration of positive solutions for a p-fractional Choquard equation in R^N. *Comput. Math. Appl.* **2019**. [CrossRef]
37. Küpper, T.; Zhang, Z.; Xia, H. Multiple positive solutions and bifurcation for an equation related to Choquard's equation. *Proc. Edinb. Math. Soc.* **2003**, *46*, 597–607. [CrossRef]
38. Shen, Z.F.; Gao, F.S.; Yang, M.B. Multiple solutions for nonhomogeneous Choquard eqution involving Hardy-Littlewood-Sobolev critical exponent. *Z. Angew. Math. Phys.* **2017**, *68*, 61. [CrossRef]
39. Cerami, G.; Vaira, G. Multiple solutions for nonhomogeneous Schrödinger-Maxwell and Klein-Gordon-Maxwell equations on R^3. *J. Differ. Equ.* **2010**, *248*, 521–543. [CrossRef]
40. Chen, S.J.; Tang, C.L. Multiple solutions for nonhomogeneous Schrödinger-Maxwell and Klein-Gordon-Maxwell equations on R^3. *Nonlinear Differ. Equ. Appl.* **2010**, *17*, 559–574. [CrossRef]
41. Wang, L. Multiple solutions for nonhomogeneous Choquard equations. *Electron. J. Differ. Equ.* **2018**, *172*, 1–27.
42. Su, Y.; Chen, H.B. Existence of nontrivial solutions for a perturbation of Choquard equation with Hardy-Littlewood-Sobolev upper critical exponent. *Electron. J. Differ. Equ.* **2018**, *2018*, 1–25.
43. Zhang, Z.J. Multiple solutions of nonhomogeneous Chouquard's equations. *Acta Math. Appl. Sin.* **2001**, *17*, 47–52.
44. Tarantello, G. On nonhomogeneous elliptic equations involving critical Sobolev exponent. In *Annales de l'Institut Henri Poincare (C) Non Linear Analysis*; Elsevier Masson: Paris, France, 1992.
45. Lin, X.; Tang, X.H. Existence of infinitely many solutions for p-Laplacian equations in R^N. *Nonlinear Anal.* **2013**, *92*, 72–81. [CrossRef]
46. Chang, K.C. *Methods in Nonlinear Analysis*; Springer: Berlin/Heidelberg, Germany, 2005.

mathematics MDPI

Article

New Generalized Mizoguchi-Takahashi's Fixed Point Theorems for Essential Distances and e^0-Metrics

Binghua Jiang [1], Huaping Huang [2] and Wei-Shih Du [3,*]

1 School of Mathematics and Statistics, Hubei Normal University, Huangshi 435002, China; jbh510@163.com
2 School of Mathematics and Statistics, Chongqing Three Gorges University, Wanzhou 404020, China; mathhhp@163.com
3 Department of Mathematics, National Kaohsiung Normal University, Kaohsiung 82444, Taiwan
* Correspondence: wsdu@mail.nknu.edu.tw

Received: 9 November 2019; Accepted: 9 December 2019; Published: 11 December 2019

Abstract: In this paper, we present some new generalizations of Mizoguchi-Takahashi's fixed point theorem which also improve and extend Du-Hung's fixed point theorem. Some new examples illustrating our results are also given. By applying our new results, some new fixed point theorems for essential distances and e^0-metrics were established.

Keywords: $\mathcal{MT}$-function; $\mathcal{MT}(\lambda)$-function; τ-function; essential distance (e-distance); e^0-metric; Du-Hung's fixed point theorem; Mizoguchi-Takahashi's fixed point theorem; Nadler's fixed point theorem; Banach contraction principle

MSC: 47H10; 54H25

1. Introduction

Let (W,ρ) be a metric space. For each $a \in W$ and any nonempty subset M of W, let

$$\rho(a,M) = \inf_{b \in M} \rho(a,b).$$

Denote by $\mathcal{N}(W)$, the family of all nonempty subsets of W, and by $\mathcal{CB}(W)$, the class of all nonempty closed and bounded subsets of W. A function $\mathcal{H} : \mathcal{CB}(W) \times \mathcal{CB}(W) \to [0,+\infty)$ defined by

$$\mathcal{H}(C,D) = \max\left\{\sup_{a \in D} \rho(a,C), \sup_{a \in C} \rho(a,D)\right\}$$

is said to be the *Hausdorff metric* on $\mathcal{CB}(W)$ induced by the metric ρ on W. A point z in W is a fixed point of a mapping T if $z = Tz$ (when $T : W \to W$ is a single-valued mapping) or $z \in Tz$ (when $T : W \to \mathcal{N}(W)$ is a multivalued mapping). The set of fixed points of T is denoted by $\mathcal{F}(T)$.

Fixed point theory is a fascinating mathematical theory that has a wide range of applications in many areas of mathematics, including nonlinear analysis, optimization, variational inequality problems, integral and differential equations and inclusions, critical point theory, nonsmooth analysis, dynamic system theory, control theory, economics, game theory, finance mathematics and so on. The famous Banach contraction principle [1] is undoubtedly one of the most important and applicable fixed point theorems which has played a significant role in nonlinear analysis and applied mathematical analysis. Many authors have devoted their attentions to study generalizations in various different directions of the Banach contraction principle; see, e.g., [2–23] and references therein.

Theorem 1. *(Banach [1]) Let (W, ρ) be a complete metric space and $T: W \to W$ be a selfmapping. Assume that there exists a nonnegative number $\lambda < 1$ such that*

$$\rho(Ta, Tb) \leq \lambda \rho(a, b) \text{ for all } a, b \in W.$$

Then T has a unique fixed point in W.

Nadler's fixed point theorem [21] was established in 1969 to extend the Banach contraction principle for multivalued mappings.

Theorem 2. *(Nadler [21]) Let (W, ρ) be a complete metric space and $T : W \to \mathcal{CB}(W)$ be a multivalued mapping. Suppose that there exists a nonnegative number $\lambda < 1$ such that*

$$\mathcal{H}(Ta, Tb) \leq \lambda \rho(a, b) \quad \text{for all } a, b \in W.$$

Then T has a fixed point in W.

Later, in 1989, Mizoguchi and Takahashi [20] presented a celebrated generalization of Nadler's fixed point theorem. In 2008, Suzuki gave an example [22] (Example 1) to show that Mizoguchi-Takahashi's fixed point theorem is a real generalization of Nadler's fixed point theorem.

Theorem 3. *(MizoguchiandTakahashi [20]) Let (W, ρ) be a complete metric space and $T : W \to \mathcal{CB}(W)$ be a multivalued mapping. Assume that*

$$\mathcal{H}(Ta, Tb) \leq \mu(\rho(a, b))\rho(a, b) \quad \text{for all } a, b \in W,$$

where μ: $[0, +\infty) \to [0, 1)$ is an $\mathcal{MT}$-function; that is, μ satisfies $\limsup_{x \to s^+} \mu(x) < 1$ for all $s \in [0, +\infty)$. Then T has a fixed point in W.

A number of generalizations of Mizoguchi-Takahashi's fixed point theorem were studied; see [2,4,8–13,15,16] and references therein. In 2016, Du and Hung [10] established the following generalized Mizoguchi-Takahashi's fixed point theorem.

Theorem 4. *(Du and Hung [10]) Let (W, ρ) be a complete metric space, $T : W \to CB(W)$ be a multivalued mapping and $\mu : [0, +\infty) \to [0, 1)$ be an $\mathcal{MT}$-function. Suppose that*

$$\min\{\mathcal{H}(Ta, Tb), \rho(a, Ta)\} \leq \mu(\rho(a, b))\rho(a, b) \quad \text{for all } a, b \in W \text{ with } a \neq b.$$

Then T admits a fixed point in W.

Theorem 4 is different from known generalizations in the existing literature and was illustrated by [7] (Example A) in which Mizoguchi-Takahashi's fixed point theorem is not applicable.

In this paper, we establish some new generalizations of Mizoguchi-Takahashi's fixed point theorem which also improve and extend Du-Hung's fixed point theorem. Some new examples illustrating our results are also given. By applying our new results, we obtained some new fixed point theorems for essential distances and e^0-metrics.

2. Preliminaries

Let (W, ρ) be a metric space. A real valued function $f : W \to \mathbb{R}$ is called *lower semicontinuous* if $\{x \in W : f(x) \leq r\}$ is *closed* for any $r \in \mathbb{R}$. Recall that a function $\kappa : W \times W \to [0, +\infty)$ is called a *w-distance* [14,18], if the following are satisfied:

($w1$) $\kappa(a,c) \leq \kappa(a,b) + \kappa(b,c)$ for any $a,b,c \in W$;
($w2$) For any $a \in W$, $\kappa(a,\cdot) : W \to [0,+\infty)$ is lower semicontinuous;
($w3$) For any $\varepsilon > 0$, there exists $\delta > 0$ such that $\kappa(c,a) \leq \delta$ and $\kappa(c,b) \leq \delta$ imply $\rho(a,b) \leq \varepsilon$.

A function $\kappa : W \times W \to [0,+\infty)$ is said to be a τ-*function* [2,3,6,8,9,17,19], if the following conditions hold:

($\tau1$) $\kappa(a,c) \leq \kappa(a,b) + \kappa(b,c)$ for any $a,b,c \in W$;
($\tau2$) If $a \in W$ and $\{b_n\}$ in W with $\lim_{n\to\infty} b_n = b$ such that $\kappa(a,b_n) \leq \beta$ for some $\beta = \beta(a) > 0$, then $\kappa(a,b) \leq \beta$;
($\tau3$) For any sequence $\{a_n\}$ in W with $\limsup_{n\to\infty}\{\kappa(a_n,a_m) : m > n\} = 0$, if there exists a sequence $\{b_n\}$ in X such that $\lim_{n\to\infty} \kappa(a_n,b_n) = 0$, then $\lim_{n\to\infty} \rho(a_n,b_n) = 0$;
($\tau4$) For $a,b,c \in W$, $\kappa(a,b) = 0$ and $\kappa(a,c) = 0$ imply $b = c$.

It is obvious that the metric ρ is a w-distance and any w-distance is a τ-function, but the converse is not true; see [2,17] for more details.

The following result is useful in our proofs.

Lemma 1. *(See [6] (Lemma 1.1).) If condition $(\tau4)$ is weakened to the following condition $(\tau4)'$:*

$(\tau4)'$ for any $a \in W$ with $\kappa(a,a) = 0$, if $\kappa(a,b) = 0$ and $\kappa(a,c) = 0$, then $b = c$,

then $(\tau3)$ implies $(\tau4)'$.

In 2016, Du [6] introduced the concept of essential distance; see also [8].

Definition 1. *(See [6] (Definition 1.2).) Let (W,d) be a metric space. A function $\kappa : W \times W \to [0,+\infty)$ is called an essential distance (abbreviated as "e-distance") if conditions $(\tau1)$, $(\tau2)$ and $(\tau3)$ hold.*

Remark 1.

(i) Clearly, any τ-function is an e-distance.
(ii) By Lemma 1, we know that if κ is an e-distance, then condition $(\tau4)'$ holds.

The following known result is crucial in this paper.

Lemma 2. *(See [3] (Lemma 2.1).) Let (W,ρ) be a metric space and $\kappa : W \times W \to [0,+\infty)$ be a function. Assume that κ satisfies the condition $(\tau3)$. If a sequence $\{a_n\}$ in W with $\lim\limits_{n\to\infty} \sup\{\kappa(a_n,a_m) : m > n\} = 0$, then $\{a_n\}$ is a Cauchy sequence in W.*

In 2016, Du introduced the concept of $\mathcal{MT}(\lambda)$-function [5] as follows (see also [7]).

Definition 2. *Let $\lambda > 0$. A function $\tau : [0,+\infty) \to [0,\lambda)$ is said to be an $\mathcal{MT}(\lambda)$-function [5] if $\limsup\limits_{x\to\gamma^+} \tau(x) < \lambda$ for all $\gamma \in [0,+\infty)$. As usual, we simply write "$\mathcal{MT}$-function" instead of "$\mathcal{MT}(1)$-function".*

A useful characterization theorem for $\mathcal{MT}(\lambda)$-functions was established by Du [5] in 2016 as follows.

Theorem 5. *(See [5] (Theorem 2.4).) Let $\lambda > 0$ and let $\tau : [0,+\infty) \to [0,\lambda)$ be a function. Then the following statements are equivalent.*

(1) τ is an $\mathcal{MT}(\lambda)$-function.

(2) $\lambda^{-1}\tau$ is an $\mathcal{MT}$-function.

(3) For each $\gamma \in [0, +\infty)$, there exists $\xi_t^{(1)} \in [0, \lambda)$ and $\epsilon_t^{(1)} > 0$ such that $\tau(x) \leq \xi_t^{(1)}$ for all $x \in (\gamma, \gamma + \epsilon_t^{(1)})$.

(4) For each $\gamma \in [0, +\infty)$, there exists $\xi_t^{(2)} \in [0, \lambda)$ and $\epsilon_t^{(2)} > 0$ such that $\tau(x) \leq \xi_t^{(2)}$ for all $x \in [\gamma, \gamma + \epsilon_t^{(2)}]$.

(5) For each $\gamma \in [0, +\infty)$, there exists $\xi_t^{(3)} \in [0, \lambda)$ and $\epsilon_t^{(3)} > 0$ such that $\tau(x) \leq \xi_t^{(3)}$ for all $x \in (\gamma, \gamma + \epsilon_t^{(3)}]$.

(6) For each $\gamma \in [0, +\infty)$, there exists $\xi_t^{(4)} \in [0, \lambda)$ and $\epsilon_t^{(4)} > 0$ such that $\tau(x) \leq \xi_t^{(4)}$ for all $x \in [\gamma, \gamma + \epsilon_t^{(4)})$.

(7) For any nonincreasing sequence $\{\beta_n\}_{n\in\mathbb{N}}$ in $[0, +\infty)$, we have $0 \leq \sup_{n\in\mathbb{N}} \tau(\beta_n) < \lambda$.

(8) For any strictly decreasing sequence $\{\beta_n\}_{n\in\mathbb{N}}$ in $[0, +\infty)$, we have $0 \leq \sup_{n\in\mathbb{N}} \tau(\beta_n) < \lambda$.

(9) For any eventually nonincreasing sequence $\{\beta_n\}_{n\in\mathbb{N}}$ (i.e., there exists $\alpha \in \mathbb{N}$ such that $\beta_{n+1} \leq \beta_n$ for all $n \in \mathbb{N}$ with $n \geq \alpha$) in $[0, +\infty)$, we have $0 \leq \sup_{n\in\mathbb{N}} \tau(\beta_n) < \lambda$.

(10) For any eventually strictly decreasing sequence $\{\beta_n\}_{n\in\mathbb{N}}$ (i.e., there exists $\alpha \in \mathbb{N}$ such that $\beta_{n+1} < \beta_n$ for all $n \in \mathbb{N}$ with $n \geq \alpha$) in $[0, +\infty)$, we have $0 \leq \sup_{n\in\mathbb{N}} \tau(\beta_n) < \lambda$.

Let κ be an e-distance on a metric space (W, ρ). For each $a \in W$ and any nonempty subset G of W, we define $\kappa(a, G)$ by

$$\kappa(a, G) = \inf_{b\in G} \kappa(a, b).$$

The following Lemma is essentially proved in [2].

Lemma 3. *(See [2] (Lemma 1.2).) Let G be a closed subset of a metric space (W, ρ) and κ be a function satisfying the condition $(\tau 3)$. Suppose that there exists $c \in W$ such that $\kappa(c, c) = 0$. Then $\kappa(c, G) = 0$ if and only if $c \in G$.*

Very recently, Du introduced and studied the concept of e^0-distance [9].

Definition 3. *(See [9] (Definition 1.3).) Let (W, ρ) be a metric space. A function $\kappa : W \times W \to [0, +\infty)$ is called an e^0-distance if it is an e-distance on W with $\kappa(a, a) = 0$ for all $a \in W$.*

Remark 2. *By applying Lemma 1, if κ is an e^0-distance on W, then for $a, b \in W$, $\kappa(a, b) = 0 \Longleftrightarrow a = b$.*

Example 1. *Let $W = \mathbb{R}$ with the metric $\rho(a, b) = |a - b|$. Then (W, ρ) is a metric space. Define the function $\kappa : W \times W \to [0, +\infty)$ by*

$$\kappa(x, y) = \max\{9(x - y), 5(y - x)\}.$$

Therefore κ is not a metric due to its asymmetry. It is easy to see that κ is an e^0-distance on W.

The following concept of e^0-metric was studied by Du in [9] which generalizes the concept of Hausdorff metric.

Definition 4. *(See [9] (Definition 1.4).) Let (W, ρ) be a metric space and κ be an e^0-distance. For any E, $F \in \mathcal{CB}(W)$, define a function $\mathcal{D}_\kappa : \mathcal{CB}(W) \times \mathcal{CB}(W) \to [0, +\infty)$ by*

$$\mathcal{D}_\kappa(E, F) = \max\{\xi_\kappa(E, F), \xi_\kappa(F, E)\},$$

where $\xi_\kappa(E, F) = \sup_{x\in E} \kappa(x, F)$, and then $\mathcal{D}_\kappa$ is said to be the e^0-metric on $\mathcal{CB}(W)$ induced by κ.

The following result presented in [9] (Theorem 1.3) is quite important in our proofs. Although its proof is similar to the proof of [2] (Theorem 1.2), we give it here for the sake of completeness and the readers convenience.

Theorem 6. *(See [9] (Theorem 1.3).) Let (W, ρ) be a metric space and $\mathcal{D}_\kappa$ be an e^0-metric defined as in Definition 4 on $\mathcal{CB}(W)$ induced by an e^0-distance κ. Then, for $E, F, G \in \mathcal{CB}(W)$, the following hold:*

(i) $\xi_\kappa(E, F) = 0 \Longleftrightarrow E \subseteq F$;
(ii) $\xi_\kappa(E, F) \le \xi_\kappa(E, G) + \xi_\kappa(G, F)$;
(iii) *Every e^0-metric $\mathcal{D}_\kappa$ is a metric on $\mathcal{CB}(W)$.*

Proof. To see (i), if $\xi_\kappa(E, F) = 0$, then $\kappa(a, F) = 0$ for all $a \in E$. By Lemma 3, we get $E \subseteq F$. Conversely, if $E \subseteq F$, by Lemma 3 again, we obtain $\xi_\kappa(E, F) = 0$ and (i) is proven. Fix $a \in E$ and $c \in G$. Then we have

$$\kappa(a, F) \le \kappa(a, b) \le \kappa(a, c) + \kappa(c, b) \quad \text{for all } b \in F,$$

which deduces

$$\kappa(a, F) \le \kappa(a, c) + \kappa(c, F).$$

So, for any $a \in E$, we obtain

$$\kappa(a, F) \le \inf\{\kappa(a, c) + \kappa(c, F) : c \in G\} \le \kappa(a, G) + \xi_\kappa(G, F).$$

Taking the supremum on both sides of the last inequality over all $a \in E$, we can obtain (ii). Finally, we verify (iii). Obviously, $\mathcal{D}_\kappa(E, F) \ge 0$ and $\mathcal{D}_\kappa(E, F) = \mathcal{D}_\kappa(F, E)$. By using (i), we have $\mathcal{D}_\kappa(E, F) = 0 \Longleftrightarrow E = F$. Applying (ii), we have

$$\begin{aligned}\mathcal{D}_\kappa(E, F) &= \max\{\xi_\kappa(E, F), \xi_\kappa(F, E)\}\\ &\le \max\{\xi_\kappa(E, G) + \xi_\kappa(G, F), \xi_\kappa(F, G) + \xi_\kappa(G, E)\}\\ &\le \mathcal{D}_\kappa(E, G) + \mathcal{D}_\kappa(G, F).\end{aligned}$$

These arguments show that $\mathcal{D}_\kappa$ is a metric on $\mathcal{CB}(W)$. □

Definition 5. *Let U be a nonempty subset of a metric space (W, ρ) and κ be an e-distance on W. A multivalued mapping $T : U \to \mathcal{N}(W)$ is said to have the κ-approximate fixed point property in U provided $\inf_{a \in U} \kappa(a, Ta) = 0$. In particular, if $\kappa \equiv \rho$, then T is said to have the approximate fixed point property in U.*

Remark 3. *Let U be a nonempty subset of a metric space (W, ρ) and $T : U \to \mathcal{N}(W)$ be a multivalued mapping. Clearly, $\mathcal{F}(T) \cap U \neq \emptyset$ implies that T has the approximate fixed point property in U.*

3. Main Results

In this section, we first prove a new generalized Mizoguchi-Takahashi's fixed point theorem with a new nonlinear condition.

Theorem 7. *Let (W, ρ) be a metric space and $\mathcal{D}_\kappa$ be an e^0-metric on $\mathcal{CB}(W)$ induced by an e^0-distance κ. Let $T : W \to \mathcal{CB}(W)$ be a multivalued mapping and $\varphi : [0, +\infty) \to [0, 1)$ be an $\mathcal{MT}$-function. Assume that*

$$\kappa(a, x) \le \kappa(x, a) \quad \text{for all } a \in Tx \tag{1}$$

and

$$\min\{\mathcal{D}_\kappa(Tu, Tv), \kappa(u, Tu)\} \le \varphi(\kappa(u, v))\kappa(u, v) \quad \text{for all } u, v \in W \text{ with } u \neq v. \tag{2}$$

Then, the following statements hold:

(a) *For any $z_0 \in W$, there exists a Cauchy sequence $\{z_n\}_{n=0}^{\infty}$ in W started at z_0 satisfying $z_n \in Tz_{n-1}$ for each $n \in \mathbb{N}$ and*

$$\lim_{n\to\infty} \kappa(z_n, z_{n-1}) = \lim_{n\to\infty} \kappa(z_{n-1}, z_n) = \inf_{n\in\mathbb{N}} \kappa(z_n, z_{n-1}) = \inf_{n\in\mathbb{N}} \kappa(z_{n-1}, z_n) = 0;$$

(b) *T has the κ-approximate fixed point property in W.*

Moreover, if W is complete and T further satisfies one of the following conditions:

(D1) *T is closed; that is, $GrT = \{(a,b) \in W \times W : b \in Ta\}$, the graph of T, is closed in $W \times W$;*
(D2) *The function $f : W \to \mathbb{R}$ defined by $f(a) = \kappa(a, Ta)$ is lower semicontinuous;*
(D3) *The function $g : W \to \mathbb{R}$ defined by $g(a) = \rho(a, Ta)$ is lower semicontinuous;*
(D4) *For each sequence $\{z_n\}$ in W with $z_{n+1} \in Tz_n$, $n \in \mathbb{N}$ and $\lim_{n\to\infty} z_n = w$, we have $\lim_{n\to\infty} \kappa(z_n, Tw) = 0$;*
(D5) *$\inf\{\kappa(a,v) + \kappa(a,Ta) : a \in W\} > 0$ for every $v \notin \mathcal{F}(T)$,*

then T admits a fixed point in W.

Proof. Let $\tau : [0,+\infty) \to [0,1)$ be defined by

$$\tau(x) = \frac{1}{2}(\varphi(x) + 1) \quad \text{for all } x \in [0,+\infty).$$

Hence $0 \le \varphi(x) < \tau(x) < 1$ for all $x \in [0,\infty)$. Given $b \in W$. Take $z_0 = b \in W$ and choose $z_1 \in Tz_0$. If $z_1 = z_0$, then $z_0 \in \mathcal{F}(T)$ and we are done. Otherwise, if $z_1 \neq z_0$, then $\kappa(z_1, z_0) > 0$ and we obtain from (2) that

$$\min\{\mathcal{D}_\kappa(Tz_1, Tz_0), \kappa(z_1, Tz_1)\} \le \varphi(\kappa(z_1,z_0))\kappa(z_1,z_0) < \tau(\kappa(z_1,z_0))\kappa(z_1,z_0). \tag{3}$$

Since

$$\kappa(z_1, Tz_1) \le \sup_{w\in Tx_0} \kappa(w, Tz_1) \le \mathcal{D}_\kappa(Tz_0, Tz_1) = \mathcal{D}_\kappa(Tz_1, Tz_0),$$

we get

$$\min\{\mathcal{D}_\kappa(Tz_1, Tz_0), \kappa(z_1, Tz_1)\} = \kappa(z_1, Tz_1). \tag{4}$$

Hence, by (3) and (4), we obtain

$$\kappa(z_1, Tz_1) < \tau(\kappa(z_1,z_0))\kappa(z_1,z_0),$$

which deduces that there exists $z_2 \in Tz_1$ such that

$$\kappa(z_1, z_2) < \tau(\kappa(z_1,z_0))\kappa(z_1,z_0).$$

Since $z_2 \in Tz_1$, by (1), we have

$$\kappa(z_2, z_1) < \tau(\kappa(z_1,z_0))\kappa(z_1,z_0).$$

Next, if $z_2 = z_1$, then $z_1 \in \mathcal{F}(T)$ and we finish the proof. Otherwise, since

$$\kappa(z_2, Tz_2) = \min\{\mathcal{D}_\kappa(Tz_2, Tz_1), \kappa(z_2, Tz_2)\} < \tau(\kappa(z_2,z_1))\kappa(z_2,z_1),$$

there exists $z_3 \in Tz_2$ such that

$$\kappa(z_2, z_3) < \tau(\kappa(z_2,z_1))\kappa(z_2,z_1).$$

By (1), we have

$$\kappa(z_3, z_2) < \tau(\kappa(z_2,z_1))\kappa(z_2,z_1).$$

So, by induction, we can obtain a sequence $\{z_n\}_{n\in\mathbb{N}\cup\{0\}}$ in W satisfying the following: for each $n \in \mathbb{N}$,

(i) $z_n \in Tz_{n-1}$ with $z_n \neq z_{n-1}$;
(ii) $\kappa(z_n, z_{n+1}) < \tau(\kappa(z_n, z_{n-1}))\kappa(z_n, z_{n-1})$;
(iii) $\kappa(z_{n+1}, z_n) < \tau(\kappa(z_n, z_{n-1}))\kappa(z_n, z_{n-1})$.

By (iii), the sequence $\{\kappa(z_n, z_{n-1})\}_{n\in\mathbb{N}}$ is strictly decreasing in $[0, +\infty)$. Hence

$$\lim_{n\to\infty} \kappa(z_n, z_{n-1}) = \inf_{n\in\mathbb{N}} \kappa(z_n, z_{n-1}) \quad \text{exists.} \tag{5}$$

Since φ is an $\mathcal{MT}$-function, by applying (8) of Theorem 5 with $\lambda = 1$, we obtain

$$0 \leq \sup_{n\in\mathbb{N}} \varphi(\kappa(z_n, z_{n-1})) < 1.$$

So we get

$$0 < \sup_{n\in\mathbb{N}} \tau(\kappa(z_n, z_{n-1})) = \frac{1}{2}\left[1 + \sup_{n\in\mathbb{N}} \varphi(\kappa(z_n, z_{n-1}))\right] < 1.$$

Put $\gamma := \sup_{n\in\mathbb{N}} \tau(\kappa(z_n, z_{n-1}))$. Thus $\gamma \in (0,1)$. For any $n \in \mathbb{N}$, by (iii) again, we have

$$\kappa(z_{n+1}, z_n) < \tau(\kappa(z_n, z_{n-1}))\kappa(z_n, z_{n-1}) \leq \gamma\kappa(z_n, z_{n-1}). \tag{6}$$

By (6), we get

$$\kappa(z_{n+1}, z_n) < \gamma\kappa(z_n, z_{n-1}) < \cdots < \gamma^n\kappa(z_1, z_0) \quad \text{for each } n \in \mathbb{N}. \tag{7}$$

Since $0 < \gamma < 1$, by taking the limit as $n \to \infty$ in (7), we obtain

$$\lim_{n\to\infty} \kappa(z_n, z_{n-1}) = 0. \tag{8}$$

Taking into account (5) and (8), we obtain

$$\lim_{n\to\infty} \kappa(z_n, z_{n-1}) = \inf_{n\in\mathbb{N}} \kappa(z_n, z_{n-1}) = 0.$$

On the other hand, from (ii) and using (1), we have

$$\kappa(z_n, z_{n+1}) < \gamma\kappa(z_n, z_{n-1}) \leq \gamma\kappa(z_{n-1}, z_n) \quad \text{for each } n \in \mathbb{N}.$$

which shows that the sequence $\{\kappa(z_{n-1}, z_n)\}_{n\in\mathbb{N}}$ is also strictly decreasing in $[0, +\infty)$, and hence, we can deduce

$$\kappa(z_n, z_{n+1}) < \gamma^n\kappa(z_0, z_1) \quad \text{for each } n \in \mathbb{N}. \tag{9}$$

So, by (9), we get

$$\lim_{n\to\infty} \kappa(z_{n-1}, z_n) = \inf_{n\in\mathbb{N}} \kappa(z_{n-1}, z_n) = 0. \tag{10}$$

Since $z_n \in Tz_{n-1}$ for all $n \in \mathbb{N}$, by (10), we prove

$$\inf_{a\in W} \kappa(a, Ta) = \inf_{n\in\mathbb{N}} \kappa(z_{n-1}, z_n) = 0;$$

that is, T has the κ-approximate fixed point property in W. Next, we claim that $\{z_n\}_{n\in\mathbb{N}\cup\{0\}}$ is a Cauchy sequence in W. For $m,n \in \mathbb{N}$ with $m > n$, we have from (9) that

$$\kappa(z_n, z_m) \leq \sum_{j=n}^{m-1} \kappa(z_j, z_{j+1}) < \frac{\gamma^n}{1-\gamma}\kappa(z_0, z_1). \tag{11}$$

Since $0 < \gamma < 1$, the last inequality implies

$$\lim_{n\to\infty} \sup\{\kappa(z_n, z_m) : m > n\} = 0. \tag{12}$$

Applying Lemma 2, we prove that $\{z_n\}_{n\in\mathbb{N}\cup\{0\}}$ is a Cauchy sequence in W.

Now, we assume that W is complete. We want to show $\mathcal{F}(T) \neq \emptyset$ if T further satisfies one of conditions (D1)–(D5). Since $\{z_n\}_{n\in\mathbb{N}\cup\{0\}}$ is Cauchy in W and W is complete, there exists $w \in W$ such that $z_m \to w$ as $m \to \infty$. From $(\tau 2)$ and (11), we have

$$\kappa(z_n, w) \leq \frac{\gamma^n}{1-\gamma}\kappa(z_0, z_1) \quad \text{for all } n \in \mathbb{N}. \tag{13}$$

In order to finish the proof, it is sufficient to show $w \in \mathcal{F}(T)$. If (D1) holds, since T is closed and $z_n \in Tz_{n-1}$ and $z_n \to w$ as $n \to \infty$, we get $w \in Tw$. If (D2) holds, by the lower semicontinuity of f, $z_n \to w$ as $n \to \infty$ and (10), we obtain

$$\begin{aligned} \kappa(w, Tw) &= f(w) \\ &\leq \liminf_{n\to\infty} \kappa(z_n, Tz_n) \\ &\leq \lim_{n\to\infty} k(z_n, z_{n+1}) = 0. \end{aligned}$$

By Lemma 3, $w \in \mathcal{F}(T)$. Suppose that (D3) is satisfied. Since $\{z_n\}$ is Cauchy, we have $\lim_{n\to\infty} \rho(z_n, z_{n+1}) = 0$. So, by the lower semicontinuity of g and $z_n \to w$ as $n \to \infty$, we get

$$\rho(w, Tw) = g(w) \leq \lim_{n\to\infty} \rho(z_n, z_{n+1}) = 0.$$

By the closedness of Tw, we show $w \in \mathcal{F}(T)$. Assume that (D4) holds. By (12), there exists $\{u_n\} \subset \{z_n\}$ with $\limsup_{n\to\infty}\{\kappa(u_n, u_m) : m > n\} = 0$ and $\{v_n\} \subset Tw$ such that $\lim_{n\to\infty} \kappa(u_n, v_n) = 0$. By $(\tau 3)$, $\lim_{n\to\infty} \rho(u_n, v_n) = 0$. Since $\rho(v_n, w) \leq \rho(v_n, u_n) + \rho(u_n, w)$, we have $v_n \to w$ as $n \to \infty$. By the closedness of Tw, we obtain $w \in Tw$. Finally, suppose that (D5) holds. If $w \notin Tw$, then, by (11) and (13), we obtain

$$\begin{aligned} 0 &< \inf_{a\in W}\{k(a, w) + k(a, Ta)\} \\ &\leq \inf_{n\in\mathbb{N}}\{k(z_n, w) + k(z_n, Tz_n)\} \\ &\leq \inf_{n\in\mathbb{N}}\{k(z_n, w) + k(z_n, z_{n+1})\} \\ &\leq \lim_{n\to\infty} \frac{2\gamma^n}{1-\gamma}\kappa(z_0, z_1) \\ &= 0, \end{aligned}$$

which leads to a contradiction. Therefore, it must be $w \in \mathcal{F}(T)$. The proof is completed. □

Here, we give a simple example illustrating Theorem 7.

Example 2. *Let $W = [0, +\infty)$ with the metric $\rho(x,y) = |x-y|$ for $x,y \in W$. Let $Tx = [0,x]$ for $x \in W$. It is obvious that each $x \in W$ is a fixed point of T. Let φ be any $\mathcal{MT}$-function. Let $\kappa : W \times W \to [0,+\infty)$ be defined by*

$$\kappa(u,v) = \max\{9(u-v), 5(v-u)\} \quad \text{for } u,v \in W.$$

Then, κ is an e^0-metric on W. Given $x \in W$. For any $a \in Tx = [0,x]$, we have

$$\kappa(a,x) = 5(x-a) \le 9(x-a) = \kappa(x,a),$$

which shows that (1) holds. Clearly, the function $x \mapsto \rho(x,Tx)$ is a zero function on W, so it is lower semicontinuous. Hence (D3) holds. We now claim

$$\min\{\mathcal{D}_\kappa(Tu,Tv), \kappa(u,Tu)\} \le \varphi(\kappa(u,v))\kappa(u,v) \quad \text{for all } u,v \in W \text{ with } u \ne v.$$

We consider the following two possible cases:

Case 1. *If $0 \le u < v$, we have*

$$\kappa(u,Tu) = 0,$$

$$\mathcal{D}_\kappa(Tu,Tv) = \max\left\{\sup_{z\in Tu} \kappa(z,Tv), \sup_{z\in Tv} \kappa(z,Tu)\right\} = 9(v-u)$$

and

$$\kappa(u,v) = 5(v-u).$$

So, $\min\{\mathcal{D}_\kappa(Tu,Tv), \kappa(u,Tu)\} = 0 \le \varphi(\kappa(u,v))\kappa(u,v)$.

Case 2. *If $0 \le v < u$, we obtain*

$$\kappa(u,Tu) = 0,$$

$$\mathcal{D}_\kappa(Tu,Tv) = 9(u-v)$$

and

$$\kappa(u,v) = 9(u-v).$$

Hence, $\min\{\mathcal{D}_\kappa(Tu,Tv), \kappa(u,Tu)\} = 0 \le \varphi(\kappa(u,v))\kappa(u,v)$.

By Cases 1 and 2, our claim is verified, and hence, (2) holds. Therefore, all the assumptions of Theorem 7 are satisfied and we also show that T has a fixed point in W from Theorem 7. Notice that

$$\mathcal{H}(T(5), T(9)) = 4 > \varphi(\rho(5,9))\rho(5,9),$$

so Mizoguchi-Takahashi's fixed point theorem is not applicable here. This example shows that Theorem 7 is a real generalization of Mizoguchi-Takahashi's fixed point theorem.

Remark 4. *Du-Hung's fixed point theorem (i.e., Theorem 4) can be proven immediately from Theorem 7. Indeed, let $\kappa \equiv \rho$. Then, (1) and (2), as in Theorem 7, are satisfied. We claim that (D4) as in Theorem 7 holds. Let $\{z_n\}$ in X with $z_{n+1} \in Tz_n$, $n \in \mathbb{N}$ and $\lim_{n\to\infty} z_n = w$. We obtain*

$$\begin{aligned}\lim_{n\to\infty} \rho(z_{n+1}, Tw) &\le \lim_{n\to\infty} \mathcal{H}(Tz_n, Tw) \\ &\le \lim_{n\to\infty} \{\varphi(\rho(z_n,w))\rho(z_n,w)\} = 0,\end{aligned}$$

which shows that (D4) holds. Therefore, all the assumptions of Theorem 7 are satisfied. By applying Theorem 7, we prove $\mathcal{F}(T) \ne \emptyset$.

In Theorem 7, if $T : W \to W$ is a self-mapping, then we obtain the following new fixed point theorem which generalizes Banach contraction principle.

Corollary 1. *Let (W, ρ) be a metric space, $T : W \to W$ be a self-mapping and $\varphi : [0, +\infty) \to [0, 1)$ be an $\mathcal{MT}$-function. Assume that*

$$\kappa(a, x) \leq \kappa(x, a) \quad \textit{for all } a \in Tx$$

and

$$\min\{\kappa(Tu, Tv), \kappa(u, Tu)\} \leq \varphi(\kappa(u, v))\kappa(u, v) \quad \textit{for all } u, v \in W \textit{ with } u \neq v.$$

Then the following statements hold:

(a) *For any $z_0 \in W$, there exists a Cauchy sequence $\{z_n\}_{n=0}^{\infty}$ in W started at z_0 satisfying $z_n = Tz_{n-1}$ for each $n \in \mathbb{N}$ and*

$$\lim_{n\to\infty} \kappa(z_n, z_{n-1}) = \lim_{n\to\infty} \kappa(z_{n-1}, z_n) = \inf_{n\in\mathbb{N}} \kappa(z_n, z_{n-1}) = \inf_{n\in\mathbb{N}} \kappa(z_{n-1}, z_n) = 0;$$

(b) *T has the κ-approximate fixed point property in W.*

Moreover, if W is complete and T further satisfies one of conditions (D1)-(D5) as in Theorem 7, then T admits a fixed point in W.

By applying Theorem 7, we establish some new fixed point theorems for e^0-metrics and e^0-distances.

Corollary 2. *Let (W, ρ) be a complete metric space and $\mathcal{D}_\kappa$ be an e^0-metric on $\mathcal{CB}(W)$ induced by an e^0-distance κ. Let $\varphi : [0, +\infty) \to [0, 1)$ be an $\mathcal{MT}$-function and $T : W \to \mathcal{CB}(W)$ be a multivalued mapping satisfying one of conditions (D1)-(D5) as in Theorem 7. Assume that*

$$\kappa(a, x) \leq \kappa(x, a) \quad \textit{for all } a \in Tx$$

and

$$\mathcal{D}_\kappa(Tu, Tu) + \kappa(u, Tu) \leq 2\varphi(\kappa(u, v))\kappa(u, v) \quad \textit{for all } u, v \in W \textit{ with } u \neq v. \tag{14}$$

Then T admits a fixed point in W.

Proof. For any $u, v \in W$ with $u \neq v$, by (14), we have

$$\min\{\mathcal{D}_\kappa(Tu, Tv), \kappa(u, Tu)\} \leq \frac{1}{2}\left(\mathcal{D}_\kappa(Tu, Tu) + \kappa(u, Tu)\right) \leq \varphi(\kappa(u, v))\kappa(u, v).$$

Hence the condition (2) in Theorem 7 holds. Therefore, the conclusion is immediate from Theorem 7. □

Corollary 3. *Let (W, ρ) be a complete metric space and $\mathcal{D}_\kappa$ be an e^0-metric on $\mathcal{CB}(W)$ induced by an e^0-distance κ. Let $\varphi : [0, +\infty) \to [0, 1)$ be an $\mathcal{MT}$-function and $T : W \to \mathcal{CB}(W)$ be a multivalued mapping satisfying one of conditions (D1)-(D5) as in Theorem 7. Assume that*

$$\kappa(a, x) \leq \kappa(x, a) \quad \textit{for all } a \in Tx$$

and

$$\sqrt{\mathcal{D}_\kappa(Tu, Tv)\kappa(u, Tu)} \leq \varphi(\kappa(u, v))\kappa(u, v) \quad \textit{for all } u, v \in W \textit{ with } u \neq v. \tag{15}$$

Then T admits a fixed point in W.

Proof. For any $u, v \in W$ with $u \neq v$, from (15), we obtain

$$\min\{\mathcal{D}_\kappa(Tu, Tv), \kappa(u, Tu)\} \leq \sqrt{\mathcal{D}_\kappa(Tu, Tv)\kappa(u, Tu)} \leq \varphi(\kappa(u, v))\kappa(u, v).$$

So the condition (2) in Theorem 7 holds. Hence, the conclusion is immediate from Theorem 7. □

In fact, we can establish a wide generalization of Corollary 2 as follows.

Corollary 4. *Let (W,ρ) be a complete metric space and $\mathcal{D}_\kappa$ be an e^0-metric on $\mathcal{CB}(W)$ induced by an e^0-distance κ. Let $\varphi : [0,+\infty) \to [0,1)$ be an $\mathcal{MT}$-function and $T : X \to \mathcal{CB}(W)$ be a multivalued mapping satisfying one of conditions (D1)-(D5) as in Theorem 7. Assume that*

$$\kappa(a,x) \le \kappa(x,a) \quad \text{for all } a \in Tx$$

and

$$\frac{s\mathcal{D}_\kappa(Tu,Tv) + t\kappa(u,Tv)}{s+t} \le \varphi(\kappa(u,v))\kappa(u,v) \quad \text{for all } u,v \in W \text{ with } u \neq v, \tag{16}$$

where $s,t \ge 0$ with $s+t>0$. Then T admits a fixed point in W.

Proof. For any $u,v \in W$ with $u \neq v$, by (16), we get

$$\min\{\mathcal{D}_\kappa(Tu,Tv), \kappa(u,Tu)\} \le \frac{s\mathcal{D}_\kappa(Tu,Tv) + t\kappa(u,Tv)}{s+t} \le \varphi(\kappa(u,v))\kappa(u,v),$$

and hence the condition (2) in Theorem 7 is satisfied. So the desired conclusion follows from Theorem 7 immediately. □

Now, we focus the following new fixed point theorem without the assumption (1) and satisfy another new condition

$$\min\{\mathcal{D}_\kappa(Tu,Tv), \kappa(v,Tv)\} \le \varphi(\kappa(u,v))\kappa(u,v) \quad \text{for all } u,v \in W \text{ with } u \neq v,$$

which is different from (2) as in Theorem 7. It is worth mentioning that this new fixed point theorem is meaningful because an e^0*-distance* is asymmetric in general.

Theorem 8. *Let (W,ρ) be a metric space and $\mathcal{D}_\kappa$ be an e^0-metric on $\mathcal{CB}(W)$ induced by an e^0-distance κ. Let $T : W \to \mathcal{CB}(W)$ be a multivalued mapping and $\varphi : [0,+\infty) \to [0,1)$ be an $\mathcal{MT}$-function. Assume that*

$$\min\{\mathcal{D}_\kappa(Tu,Tv), \kappa(v,Tv)\} \le \varphi(\kappa(u,v))\kappa(u,v) \quad \text{for all } u,v \in W \text{ with } u \neq v. \tag{17}$$

Then the following statements hold:

(a) For any $z_0 \in W$, there exists a Cauchy sequence $\{z_n\}_{n=0}^{\infty}$ in W started at z_0 satisfying $z_n \in Tz_{n-1}$ for each $n \in \mathbb{N}$ and

$$\lim_{n\to\infty} \kappa(z_{n-1},z_n) = \inf_{n\in\mathbb{N}} \kappa(z_{n-1},z_n) = 0;$$

(b) T has the κ-approximate fixed point property in W.

Moreover, if W is complete and T further satisfies one of conditions (D1)-(D5) as in Theorem 7, then $\mathcal{F}(T) \neq \emptyset$.

Proof. Define $\tau(x) = \frac{1}{2}(\varphi(x)+1)$ for all $x \in [0,+\infty)$. Then $0 \le \varphi(x) < \tau(x) < 1$ for all $x \in [0,+\infty)$. Let $b \in W$. Take $z_0 = b \in W$ and choose $z_1 \in Tz_0$. If $z_1 = z_0$, then $z_0 \in \mathcal{F}(T)$ and we are done. Otherwise, if $z_1 \neq z_0$, then $\kappa(z_0,z_1) > 0$. By (17), we have

$$\begin{aligned}\kappa(z_1,Tz_1) &= \min\{\mathcal{D}_\kappa(Tz_0,Tz_1), \kappa(z_1,Tz_1)\}\\ &\le \varphi(\kappa(z_0,z_1))\kappa(z_0,z_1)\\ &< \tau(\kappa(z_0,z_1))\kappa(z_0,z_1),\end{aligned}$$

from which one can deduce that there exists $z_2 \in Tz_1$ such that

$$\kappa(z_1, z_2) < \tau(\kappa(z_0, z_1))\kappa(z_0, z_1).$$

Next, if $z_2 = z_1$, then $z_1 \in \mathcal{F}(T)$, and we finish the proof. Otherwise, since

$$\kappa(z_2, Tz_2) = \min\{\mathcal{D}_\kappa(Tz_1, Tz_2), \kappa(z_2, Tz_2)\} < \tau(\kappa(z_1, z_2))\kappa(z_1, z_2),$$

then there exists $z_3 \in Tz_2$ such that

$$\kappa(z_2, z_3) < \tau(\kappa(z_1, z_2))\kappa(z_1, z_2).$$

Hence, by induction, we can obtain a sequence $\{z_n\}_{n\in\mathbb{N}\cup\{0\}}$ satisfying the following: for each $n \in \mathbb{N}$,

(iv) $z_n \in Tz_{n-1}$ with $z_n \neq z_{n-1}$;
(v) $\kappa(z_n, z_{n+1}) < \tau(\kappa(z_{n-1}, z_n))\kappa(z_{n-1}, z_n)$.

By (v), the sequence $\{\kappa(z_{n-1}, z_n)\}_{n\in\mathbb{N}}$ is strictly decreasing in $[0, +\infty)$. So

$$\lim_{n\to\infty} \kappa(z_{n-1}, z_n) = \inf_{n\in\mathbb{N}} \kappa(z_{n-1}, z_n) \quad \text{exists.} \tag{18}$$

Since φ is an $\mathcal{MT}$-function, by applying (8) of Theorem 5 with $\lambda = 1$, we obtain

$$0 \leq \sup_{n\in\mathbb{N}} \varphi(\kappa(z_{n-1}, z_n)) < 1.$$

So we get

$$0 < \sup_{n\in\mathbb{N}} \tau(\kappa(z_{n-1}, z_n)) = \frac{1}{2}\left[1 + \sup_{n\in\mathbb{N}} \varphi(\kappa(z_{n-1}, z_n))\right] < 1.$$

Hence $c := \sup_{n\in\mathbb{N}} \tau(\kappa(z_{n-1}, z_n)) \in (0, 1)$. For any $n \in \mathbb{N}$, by (v) again, we obtain

$$\kappa(z_n, z_{n+1}) < \tau(\kappa(z_{n-1}, z_n))\kappa(z_{n-1}, z_n) \leq c\kappa(z_{n-1}, z_n).$$

which implies

$$\kappa(z_n, z_{n+1}) < c^n\kappa(z_0, z_1) \quad \text{for each } n \in \mathbb{N}. \tag{19}$$

Since $0 < c < 1$, by taking the limit as $n \to \infty$ in (19), we have

$$\lim_{n\to\infty} \kappa(z_n, z_{n+1}) = 0. \tag{20}$$

Combining (18) and (20), we obtain

$$\lim_{n\to\infty} \kappa(z_{n-1}, z_n) = \inf_{n\in\mathbb{N}} \kappa(z_{n-1}, z_n) = 0 \tag{21}$$

and hence (a) is proven. To see (b), since $z_n \in Tz_{n-1}$ for all $n \in \mathbb{N}$, by (21), we show that

$$\inf_{a\in W} \kappa(a, Ta) = \inf_{n\in\mathbb{N}} \kappa(z_{n-1}, z_n) = 0.$$

Using a similar argument as in the proof of Theorem 7, one can verify that $\mathcal{F}(T) \neq \emptyset$ and finish this proof. □

The following example not only illustrates Theorem 8 but also shows that Theorem 8 is different from Theorem 7.

Example 3. *Let $W = [0, +\infty)$ with the metric $\rho(x, y) = |x - y|$ for $x, y \in W$. Let $Tx = [0, x]$ for $x \in W$. So each $x \in W$ is a fixed point of T. Let φ be any $\mathcal{MT}$-function. Let $\kappa : W \times W \to [0, +\infty)$ be defined by*

$$\kappa(u, v) = \max\{4(u - v), 7(v - u)\} \quad \text{for } u, v \in W.$$

Then κ is an e^0-metric on W. Clearly, the function $x \mapsto \rho(x, Tx)$ is a zero function on W, so it is lower and semicontinuous. Hence, (D3) holds. Using a similar argument as in Example 2, we can prove that

$$\min\{\mathcal{D}_\kappa(Tu, Tv), \kappa(v, Tv)\} \leq \varphi(\kappa(u, v))\kappa(u, v) \quad \text{for all } u, v \in W \text{ with } u \neq v.$$

Hence, all the assumptions of Theorem 8 are satisfied. Applying Theorem 8, we also prove that T has a fixed point in W. Notice that $1 \in T(2) = [0, 2]$ and

$$\kappa(1, 2) = 7 > 4 = \kappa(2, 1),$$

so (1) does not hold and hence Theorem 7 is not applicable here. Moreover, since

$$\mathcal{H}(T(3), T(8)) = 5 > \varphi(\rho(3, 8))\rho(3, 8),$$

Mizoguchi-Takahashi's fixed point theorem is also not applicable.

Some new fixed point theorems are established by Theorem 8 immediately.

Corollary 5. *Let (W, ρ) be a metric space, $T : W \to W$ be a selfmapping and $\varphi : [0, +\infty) \to [0, 1)$ be an $\mathcal{MT}$-function. Assume that*

$$\min\{\kappa(Tu, Tv), \kappa(v, Tv)\} \leq \varphi(\kappa(u, v))\kappa(u, v) \quad \text{for all } u, v \in W \text{ with } u \neq v.$$

Then the following statements hold:

(a) For any $z_0 \in W$, there exists a Cauchy sequence $\{z_n\}_{n=0}^{\infty}$ in W started at z_0 satisfying $z_n = Tz_{n-1}$ for each $n \in \mathbb{N}$ and

$$\lim_{n\to\infty} \kappa(z_{n-1}, z_n) = \inf_{n\in\mathbb{N}} \kappa(z_{n-1}, z_n) = 0;$$

(b) T has the κ-approximate fixed point property in W.

Moreover, if W is complete and T further satisfies one of conditions (D1)-(D5) as in Theorem 7, then T admits a fixed point in W.

Corollary 6. *Let (W, ρ) be a complete metric space and $\mathcal{D}_\kappa$ be an e^0-metric on $\mathcal{CB}(W)$ induced by an e^0-distance κ. Let $\varphi : [0, +\infty) \to [0, 1)$ be an $\mathcal{MT}$-function and $T : W \to \mathcal{CB}(W)$ be a multivalued mapping satisfying one of conditions (D1)–(D5) as in Theorem 7. Assume that*

$$\mathcal{D}_\kappa(Tu, Tv) + \kappa(v, Tv) \leq 2\varphi(\kappa(u, v))\kappa(u, v) \quad \text{for all } u, v \in W \text{ with } u \neq v.$$

Then $\mathcal{F}(T) \neq \emptyset$.

Corollary 7. *Let (W, ρ) be a complete metric space and $\mathcal{D}_\kappa$ be an e^0-metric on $\mathcal{CB}(W)$ induced by an e^0-distance κ. Let $\varphi : [0, +\infty) \to [0, 1)$ be an $\mathcal{MT}$-function and $T : W \to \mathcal{CB}(W)$ be a multivalued mapping satisfying one of conditions (D1)-(D5) as in Theorem 7. Assume that*

$$\sqrt{\mathcal{D}_\kappa(Tu, Tv)\kappa(v, Tv)} \leq \varphi(\kappa(u, v))\kappa(u, v) \quad \text{for all } u, v \in W \text{ with } u \neq v.$$

Then $\mathcal{F}(T) \neq \emptyset$.

Corollary 8. *Let* (W, ρ) *be a complete metric space and* $\mathcal{D}_\kappa$ *be an* e^0*-metric on* $\mathcal{CB}(W)$ *induced by an* e^0*-distance* κ*. Let* $\varphi : [0, +\infty) \to [0, 1)$ *be an* $\mathcal{MT}$*-function and* $T : W \to \mathcal{CB}(W)$ *be a multivalued mapping satisfying one of conditions (D1)-(D5) as in Theorem 7. Assume that*

$$\frac{s\mathcal{D}_\kappa(Tu, Tv) + t\kappa(v, Tv)}{s + t} \leq \varphi(\kappa(u, v))\kappa(u, v) \quad \text{for all } u, v \in W \text{ with } u \neq v,$$

where $s, t \geq 0$ *with* $s + t > 0$*. Then* $\mathcal{F}(T) \neq \emptyset$*.*

Remark 5.

(a) *Theorem 7, Corollary 4, Theorem 8 and Corollary 8 all generalize and extend Mizoguchi-Takahashi's fixed point theorem;*

(b) *All results in [10] are special cases of our results established in this paper.*

(c) *Theorems 7 and 8 improve and generalize some of the existence results on the topic in the literature; see, e.g., [1,2,4,7,8,10,11,13–16,20–23] and references therein.*

4. Conclusions

Our main purpose in this paper is to establish new generalizations of Mizoguchi-Takahashi's fixed point theorem for essential distances and e^0-metrics satisfying the following new conditions:

- $\min\{\mathcal{D}_\kappa(Tu, Tv), \kappa(u, Tu)\} \leq \varphi(\kappa(u, v))\kappa(u, v)$ for all $u, v \in W$ with $u \neq v$ (see Theorem 7 for details),
- $\min\{\mathcal{D}_\kappa(Tu, Tv), \kappa(v, Tv)\} \leq \varphi(\kappa(u, v))\kappa(u, v)$ for all $u, v \in W$ with $u \neq v$ (see Theorem 8 for details).

We give new examples to illustrate our results. As applications, some new fixed point theorems for essential distances and e^0-metrics are also established by applying these new generalized Mizoguchi-Takahashi's fixed point theorems. Our new results generalize and improve some of known results on the topic in the literature.

Author Contributions: All authors contributed equally to this work. All authors read and approved the final manuscript.

Funding: The third author is supported by grant number MOST 107-2115-M-017-004-MY2 of the Ministry of Science and Technology of the Republic of China.

Acknowledgments: The authors wish to express their hearty thanks to the anonymous referees for their valuable suggestions and comments.

Conflicts of Interest: The authors declare no conflict of interest.

References

1. Banach, S. Sur les opérations dans les ensembles abstraits et leurs applications aux équations intégrales. *Fund. Math.* **1922**, *3*, 133–181. [CrossRef]
2. Du, W.-S. Some new results and generalizations in metric fixed point theory. *Nonlinear Anal.* **2010**, *73*, 1439–1446. [CrossRef]
3. Du, W.-S. Critical point theorems for nonlinear dynamical systems and their applications. *Fixed Point Theory Appl.* **2010**, 246382. [CrossRef]
4. Du, W.-S. On coincidence point and fixed point theorems for nonlinear multivalued maps. *Topol. Appl.* **2012**, *159*, 49–56. [CrossRef]
5. Du, W.-S. New existence results of best proximity points and fixed points for $MT(\lambda)$-functions with applications to differential equations. *Linear Nonlinear Anal.* **2016**, *2*, 199–213.
6. Du, W.-S. On generalized Caristi's fixed point theorem and its equivalence. *Nonlinear Anal. Differ. Equ.* **2016**, *4*, 635–644. [CrossRef]

7. Du, W.-S. Simultaneous generalizations of fixed point theorems of Mizoguchi-Takahashi type, Nadler type Banach type, Kannan type and Chatterjea type. *Nonlinear Anal. Differ. Equ.* **2017**, *5*, 171–180. [CrossRef]
8. Du, W.-S. Some generalizations of fixed point theorems of Caristi type and Mizoguchi-Takahashi type under relaxed conditions. *Bull. Braz. Math. Soc. New Ser.* **2019**, *50*, 603–624. [CrossRef]
9. Du, W.-S. New coincidence point and fixed point theorems for essential distances and e^0-*metrics*. *arXiv* **2019**, arXiv:1907.06236v1.
10. Du, W.-S.; Hung, Y.-L. A generalization of Mizoguchi-Takahashi's fixed point theorem and its applications to fixed point theory. *Int. J. Math. Anal.* **2017**, *11*, 151–161. [CrossRef]
11. Goebel, K.; Kirk, W.A. *Topics in Metric Fixed Point Theory*; Cambridge University Press: Cambridge, UK, 1990.
12. Huang, H.; Radenović, S.; Dorixcx, D. A note on the equivalence of some metric and H-cone metric fixed point theorems for multivalued contractions. *Fixed Point Theory Appl.* **2015**, *2015*, 43. [CrossRef]
13. Hyers, D.H.; Isac, G.; Rassias, T.M. *Topics in Nonlinear Analysis and Applications*; World Scientific Publ. Co.: Singapore; Hackensack, NJ, USA; London, UK, 1997.
14. Kada, O.; Suzuki, T.; Takahashi, W. Nonconvex minimization theorems and fixed point theorems in complete metric spaces. *Math. Jpn.* **1996**, *44*, 381–391.
15. Khamsi, M.A.; Kirk, W.A. *An Introduction to Metric Spaces and Fixed Point Theory*; Wiley-Interscience: New York, NY, USA, 2001.
16. Kirk, W.A.; Shahzad, N. *Fixed Point Theory in Distance Spaces*; Springer: Cham, Switzerland, 2014.
17. Lin, L.-J.; Du, W.-S. Ekeland's variational principle, minimax theorems and existence of nonconvex equilibria in complete metric spaces. *J. Math. Anal. Appl.* **2006**, *323*, 360–370. [CrossRef]
18. Lin, L.-J.; Du, W.-S. Some equivalent formulations of generalized Ekeland's variational principle and their applications. *Nonlinear Anal.* **2007**, *67*, 187–199. [CrossRef]
19. Lin, L.-J.; Du, W.-S. On maximal element theorems, variants of Ekeland's variational principle and their applications. *Nonlinear Anal.* **2008**, *68*, 1246–1262. [CrossRef]
20. Mizoguchi, N.; Takahashi, W. Fixed point theorems for multivalued mappings on complete metric spaces. *J. Math. Anal. Appl.* **1989**, *141*, 177–188. [CrossRef]
21. Nadler, S.B., Jr. Multi-valued contraction mappings. *Pac. J. Math.* **1969**, *30*, 475–488. [CrossRef]
22. Suzuki, T. Mizoguchi-Takahashi's fixed point theorem is a real generalization of Nadler's. *J. Math. Anal. Appl.* **2008**, *340*, 752–755. [CrossRef]
23. Takahashi, W. *Nonlinear Functional Analysis*; Yokohama Publishers: Yokohama, Japan, 2000.

mathematics

MDPI

Article

Design of Robust Trackers and Unknown Nonlinear Perturbation Estimators for a Class of Nonlinear Systems: HTRDNA Algorithm for Tracker Optimization

Jiunn-Shiou Fang [1], Jason Sheng-Hong Tsai [1], Jun-Juh Yan [2,*], Chang-He Tzou [1] and Shu-Mei Guo [3]

1 Department of Electrical Engineering, National Cheng-Kung University, Tainan 701, Taiwan; fjshow611@gmail.com (J.-S.F.); shtsai@mail.ncku.edu.tw (J.S.-H.T.); patrick09091994@gmail.com (C.-H.T.)
2 Department of Electronic Engineering, National Chin-Yi University of Technology, Taichung 41107, Taiwan
3 Department of Computer Science and Information Engineering, National Cheng-Kung University, Tainan 701, Taiwan; guosm@mail.ncku.edu.tw
* Correspondence: jjyan@ncut.edu.tw

Received: 21 September 2019; Accepted: 20 November 2019; Published: 22 November 2019

Abstract: A robust linear quadratic analog tracker (LQAT) consisting of proportional-integral-derivative (PID) controller, sliding mode control (SMC), and perturbation estimator is proposed for a class of nonlinear systems with unknown nonlinear perturbation and direct feed-through term. Since the derivative type (D-type) controller is very sensitive to the state varying, a new D-type controller design algorithm is developed to avoid an unreasonable large value of the controller gain. Moreover, the boundary of D-type controller is discussed. To cope with the unknown perturbation effect, SMC is utilized. Based on the fast response of SMC controlled systems, the proposed perturbation estimator can estimate unknown nonlinear perturbation and improve the tracking performance. Furthermore, in order to tune the PID controller gains in the designed tracker, the nonlinear perturbation is eliminated by the SMC-based perturbation estimator first, then a hybrid Taguchi real coded DNA (HTRDNA) algorithm is newly proposed for the PID controller optimization. Compared with traditional DNA, a new HTRDNA is developed to improve the convergence performance and effectiveness. Numerical simulations are given to demonstrate the performance of the proposed method.

Keywords: PID controller; sliding mode control; hybrid Taguchi real coded DNA algorithm; perturbation estimator

1. Introduction

As well known, the PID controller is one of the popular control strategies and widely adopted to control engineering due to its simple structure and robust feature [1–3]. Hence, the PID controller has been widely implemented in many industrial applications. For tuning the PID controller gains, the traditional method Ziegler–Nichols rule is developed, but it is difficult to adjust the optimal or near optimal PID controller gains when the controlled system is with nonlinearity and high order dimension [3,4]. Paper [5] proposes the closed-loop controlled system by using a state-derivative feedback controller, and it illustrates the difficulty of calculating the controller based on the state-feedback control approach; hence, this paper transforms the single input single output (SISO) system into Frobenius canonical form and the pole-placement method is employed to cope with the state-derivative feedback control problem. Research work [6] processes the state-derivative feedback controller design by transforming the state-derivative feedback control problem to state-feedback control problem, but the limitation is that the system matrix A is invertible. Recently, the linear matrix inequality (LMI) approach is adopted to achieve the PID controller design. For example, the work in [7] deals with the PID controller design for the controlled system without a direct feed-through term and the output variable transformation method is adopted, but if the controlled system is with a direct feed-through term, the PID controller will become difficult to design. The authors of [8] discussed the robust PID controller for the linear uncertain system by LMI and D-stability approach. The singular system structure is used to calculate the PD controller with the H_∞ performance [9]; the H_∞ PD/PI controller design is presented in [10]. Compared with the literature in [9,10], the proposed design algorithm of PID controller is without additional singular structure. However, this paper discusses the PID-type controller in detail. For instance, the D-type controller is discussed to be bounded by a selected parameter, and the parameter is bounded in a range (0, 1); hence, the D-type controller can avoid unreasonable gain value (large gain value) through a simple proposed method.

The Laplace transform method and the final-value theorem are employed to design the tracking controller [11,12]. To shape the tracking performance, the literature in [13,14] designed the augmented state for PID filter then the controlled system is transformed to the augmented controlled system with a direct feed-through term. Moreover, the disturbance observer and functional observer are developed to measure the external disturbance [13–15]. However, the proposed design approaches [13,14] cannot be directly applied to the systems with a direct feed-through term and unknown nonlinear perturbation; hence, the PID controller is worth being developed, especially if the controlled system is with nonlinear perturbations and direct feed-through term. With the design of the PI-type controller, the controlled system has the augmented structure, and this structure may result in an uncontrollable augmented controlled system. In paper [16], the authors present a method which is placed in the closed-loop system eigenvalues on the left of the negative vertical that lies by the selected non-positive constant; hence, the proposed method is utilized to overcome the uncontrollable issue in this paper. Since the forward gain cannot be designed by using the traditional LQAT approach due to the method in [16], therefore, the final-value theorem can be adopted to overcome this problem by discussing the final-value theorem for the proposed robust tracker design in this paper.

SMC is inherently robust to external disturbance and nonlinear system and with fast response. In [17], the adaptive robust PID controller with SMC is proposed for the uncertain chaotic system. In [18], the fuzzy sliding mode control is designed for induction machine. The work in [19] designs an adaptive integral SMC for the system with uncertainty and applies the controller to the vertical take-off and landing (VTOL) aircraft system. Therefore, the SMC can be successfully utilized in many applications. Suppressing disturbance is the main target of SMC, but it cannot eliminate disturbance completely. Some researches utilize the disturbance estimators to overcome external disturbance [20,21]; the papers develop SMC to integrate with the disturbance estimator for the controlled system with undesired disturbance [22–25]. The authors of [25] propose the observer-based SMC for the controlled system with external disturbances. A robust SMC and disturbance observer via the augmented state for the multi-axis coordinated motion system is studied [26]. However, in our knowledge, the SMC-based LQAT integrated with PID controller has not been well discussed, especially if the controlled system is with a direct feed-through term. To deal with the external perturbation, this paper develops the perturbation estimator design based on the SMC due to its fast response.

The three PID controller gains must be determined properly; otherwise, it might result in undesirable performance. In the works of [27,28], the authors developed an optimization method for the PID controller design subjected to the expected performance index though the frequency response. In the work of [29], the authors proposed a methodology for designing the controller and the loop shaping with the standard performance such as H_2 and H_∞ performance. However, these proposed methodologies do not take the disturbance estimator into account [27–29]. To improve the tracking performance and control force, the disturbance estimator is adopted to the proposed controller. Recently, many popular heuristic algorithms have been applied in optimization problems. Particle swarm optimization (PSO) [3,4,30], DNA algorithm [31,32], and genetic algorithm (GA) [33–38] are stochastic searching methods for solving optimal problems. For example, some works in [33–38] based on the GA method integrated their research to the proposed controller and parameters optimization; in papers [31,32], the DNA algorithm is proposed for the PID controller optimization, and the difference between GA and DNA algorithms is the mutation operator which is not only with the same mutation operator but also consists of enzyme and virus, whereby the different PID structure can exchange their information. On the other hand, the Taguchi method is a low cost and high effective method for quality engineering [39,40]. Compared with full factorial experiments, the Taguchi method is a simple experimental design method that is less experiment. It emphasizes and focuses on the improvement of product quality not through testing but through design. Some papers apply the Taguchi method to improve the performance of GA [33,34]. Paper [33] mentions that the hybrid Taguchi–genetic algorithm (HTGA) has a quick convergent. Among the above methods, the DNA algorithm is a multiple functional method which is not only adjusted to the parameters but also changed the PID structure to find the optimal or near-optimal solution. Thus, this paper utilizes the advantage of Taguchi method to enhance the efficiency for our proposed algorithm.

Based on the above description, this paper aims to design a robust LQAT consisting of PID controller, SMC, and perturbation estimator for a class of nonlinear systems with unknown nonlinear perturbation, and the proposed HTRDNA algorithm is designed for the PID controller optimization. To avoid unreasonable gain value in the controller, a simple algorithm for D-type controller design is studied. Due to the SMC fast response, the perturbation estimator is proposed based on SMC. Since the undesirable nonlinear perturbation is eliminated by the SMC-based perturbation estimator first, it becomes easy to optimize the PID controller with the new design procedure of HTRDNA algorithm proposed in this paper.

This paper is organized as follows. Section 2 presents the whole derivation for the robust tracker design. Section 3 proposes the design procedure of HTRDNA algorithm. The illustrative examples demonstrate the feasibility and validity of the proposed approaches in Section 4 and a conclusion is given in Section 5.

Notations: w^T is used to denote the transpose for the matrix w, $w^\dagger$ denotes the matrix generalized inverse for the matrix w and $\| w \|$ denotes the Euclidean norm of the matrix w or vector w. $| w |$ represents the absolute value of w. I_n is the $n \times n$ identity matrix. $sign(s)$ is the $sign$ function of s, if $s > 0$, $sign(s) = 1$; if $s < 0$, $sign(s) = -1$; if $s = 0$, $sign(s) = 0$.

2. Robust Tracker and Perturbation Estimator Design

For a class of nonlinear systems with a direct feed-through term, the robust tracker and perturbation estimator are proposed. In real engineering systems, there are many controlled systems with nonlinear vector and disturbances such as the chaotic systems and robotic systems. To cope with these undesired perturbations, the SMC-based perturbation estimator is proposed. Now, consider a class of nonlinear time-invariant system described by

$$\dot{x}(t) = Ax(t) + B(u(t) + g(x,t) + d(x,t)), \tag{1}$$

$$y(t) = Cx(t) + Du(t), \tag{2}$$

where $A \in \Re^{n\times n}$, $B \in \Re^{n\times m}$, $C \in \Re^{p\times n}$, and $D \in \Re^{p\times m}$ denote the system matrices. The pair $(A,\ B)$ is controllable. In order to deal with the LQAT problem, the condition $m \geq p$ has to satisfy. $x(t) \in \Re^n$ is the state vector, $u(t) \in \Re^m$ is the control input, $g(x,t) \in \Re^m$ is the system nonlinear vector, and $y(t) \in \Re^p$ is the measurable output of the system. $d(x,t) \in \Re^m$ is the unknown nonlinear perturbation at time t. Notices that the proposed approach still works for the special case where $y(t) = Cx(t)$ (such as chaotic systems). Moreover, $u(t) = u^*(t) + K_D\dot{x}(t)$ where the gain K_D is D-type controller gain.

In [5,8], the closed-loop controlled system of D-type controller is discussed. Therefore, the linear transformation can be founded. To merge the derivative term $\dot{x}(t)$ in (1), theoretically it can be written to

$$(I_n - BK_D)\dot{x}(t) = Ax(t) + B(u^*(t) + g(x,t) + d(x,t)). \tag{3}$$

After being transformed, (1) can be rewritten to the following state space equation

$$\dot{x}(t) = A_{pid}x(t) + B_{pid}\big(u^*(t) + d_g(x,t)\big), \tag{4}$$

$$y(t) = C_{pid}x(t) + D_{pid1}u^*(t) + D_{pid2}d_g(x,t), \tag{5}$$

where $M = I_n - BK_D$, $A_{pid} = M^{-1}A$, $B_{pid} = M^{-1}B$, $C_{pid} = C + DK_DM^{-1}A$, $D_{pid1} = D + DK_DM^{-1}B$, $D_{pid2} = DK_DM^{-1}B$, and $d_g(x,t) = g(x,t) + d(x,t)$.

To avoid the D-type controller K_D with unreasonable values, the gain should be discussed and selected properly. In order to keep the original system feature, let the matrix M be $M = I_n - BK_D \geq \alpha I_n > 0$ where parameter α is positive definite so that the transformed system can remain its stability. Therefore, a simple D-type controller algorithm is proposed. Since the rank of BK_D is m so that $I_n - BK_D$ only m poles can be placed, some methods can be utilized to deal with this problem such as pole-placement and LMI approach. To implement minimal parameters, one solution of K_D can be obtained by

$$K_D = (1-\alpha)B^{\dagger}, \tag{6}$$

then, the matrix M is equivalent to

$$M = I_n - (1-\alpha)BB^{\dagger} > 0, \tag{7}$$

which implies

$$I_n > (1-\alpha)BB^{\dagger}. \tag{8}$$

To find out the range of α, we take 2 norm for both sides of (8)

$$||I_n|| > (1-\alpha)||BB^{\dagger}|| = (1-\alpha), \tag{9}$$

and the parameter α has the range $0 < \alpha \leq 1$. Moreover, for the requirement of the transformed matrix M being invertible. In (7)–(9), we assume that the rank of B is m, and $BB^{\dagger}$ is positive definite so that K_D should be a reasonable matrix with $0 < \alpha \leq 1$. From Equation (9), the system matrix B and $B^{\dagger}$ can be described in the singular value decomposition (SVD) form as

$$B = U_r \sum\nolimits_r V_r^T \text{ and } B^{\dagger} = V_r \sum\nolimits_r^{-1} U_r^T,$$

where $U_r \in \Re^{n \times r}$ is a unitary matrix, $\sum_r \in \Re^{r \times r}$ is the matrix with r singular values, and $V_r \in \Re^{r \times m}$ is a unitary matrix. One has

$$\begin{aligned} ||BB^{\dagger}|| &= ||U_r \Sigma_r V_r^T V_r \Sigma_r^{-1} U_r^T|| \\ &= ||U_r I_r U_r^T|| = 1. \end{aligned}$$

For the above calculation, the inverse of matrix M exists, thus, we can ensure that the transformed matrix is invertible for the linear transformation in our proposed method.

Remark 1. *If the D-type controller (6) satisfies the above design algorithm, then invertible matrix M can be computed. Since the D-type controller is sensitive to the system states varying, the gain should be selected properly. If the gain* K_D *is with the high gain property, then the* K_p *and* K_I *gains (to be appear later) will be unreasonable large. Therefore, a simple D-type controller algorithm is important.*

To construct an augmented matrix with PI-type controller. Let

$$\eta(t) = \begin{bmatrix} x(t) \\ \int e_y(t)dt \end{bmatrix}$$

to be the new state variable in the modified state space equation, where

$$e_y(t) = y(t) - r(t) \tag{10}$$

denotes the tracking error and $r(t)$ is the reference trajectory. In light of the new state variable, the system in (4) and (5) can be arranged to the new state-space equation described as

$$\dot{\eta}(t) = \overline{A}_{pid}\eta(t) + \overline{B}_{pid1}u^*(t) + \overline{B}_{pid2}d_g(x,t) - r_{pid}(t), \tag{11}$$

$$y(t) = \overline{C}_{pid}\eta(t) + \overline{D}_{pid1}u^*(t) + \overline{D}_{pid2}d_g(x,t), \tag{12}$$

where $\overline{A}_{pid} = \begin{bmatrix} A_{pid} & 0 \\ C_{pid} & 0 \end{bmatrix}$, $\overline{B}_{pid1} = \begin{bmatrix} B_{pid} \\ D_{pid1} \end{bmatrix}$, $\overline{B}_{pid2} = \begin{bmatrix} B_{pid} \\ D_{pid2} \end{bmatrix}$, $\overline{C}_{pid} = \begin{bmatrix} C_{pid} & 0 \end{bmatrix}$, $\overline{D}_{pid1} = D_{pid1}$, $\overline{D}_{pid2} = D_{pid2}$ and $r_{pid}(t) = \begin{bmatrix} 0 \\ r(t) \end{bmatrix}$. We give a sliding surface as

$$s(t) = C_s\eta(t) - \int_0^t \left(C_s\overline{A}_{pid}\eta(t) - K\eta(t) + \overline{u}(t)\right)dt, \tag{13}$$

where

$$C_s = \begin{bmatrix} B_{pid}{}^{+} & 0 \end{bmatrix}, \tag{14}$$

the equivalent control $u^*_{eq}(t)$ in the sliding manifold $\left(\dot{s}(t) = 0\right)$ is obtained by

$$u^*_{eq}(t) = -K\eta(t) + \overline{u}(t) - d_g(x,t). \tag{15}$$

We lack of the information of perturbation $d_g(x,t)$; hence, the underdetermined estimation of $d_g(x,t)$ named by $\hat{d}_g(t)$ will be design first, then the PI-type controller gain K and control law $\overline{u}(t)$ will be discussed in detail later, respectively.

Lemma 1. *In the works [15,21], the authors indicate that the perturbation is assumed to be slowly time-varying; therefore, the derivative of perturbation equal is (near) to zero. Generally, it is reasonable to suppose that*

$$\dot{d}_g(x,t) = 0, \tag{16}$$

when the perturbation is slowly time-varying and changes slightly relative to the observer dynamics with high gain property.

Give the perturbation estimator as

$$\hat{d}_g(t) = k_o\left(s(t) + \int (\gamma s(t) + \sigma sat(s(t)))dt\right), \tag{17}$$

where k_o is the positive parameter for the perturbation estimator. In the control law (15), the nonlinear perturbation $d_g(x,t)$ is unknown so that the control law cannot be achieved. Therefore, the perturbation estimator (17) can be utilized to replace the unknown nonlinear perturbation $d_g(x,t)$. Now, the SMC controller $u_{\pm}(t)$ and SMC-based control law can be designed by

$$u_{\pm}(t) = -\gamma s(t) - \sigma sat(s(t)), \tag{18}$$

$$u^*(t) = -K\eta(t) - \hat{d}_g(t) + u_{\pm}(t) + \overline{u}(t), \tag{19}$$

where γ and σ denote arbitrary nonnegative value so that the trajectories of SMC converge to the sliding manifold and the unknown nonlinear perturbation is estimated consequently.

Theorem 1. *The estimation in (17) leads to the error between the external perturbation and the estimated perturbation converge to zero closely, which implies*

$$\tilde{d}_g(t) = d_g(x,t) - \hat{d}_g(t) \approx 0. \tag{20}$$

Proof. See Appendix A. □

Remark 2. *To avoid the undesired chattering phenomenon in the SMC, the sign function can be replaced by a smooth and continuous saturation function [41].*

$$sat(s(t)) = \begin{bmatrix} \frac{s_1(t)}{|s_1(t)|+\delta_1} & \cdots & \frac{s_i(t)}{|s_i(t)|+\delta_i} \end{bmatrix}^T, \tag{21}$$

where δ_i is an arbitrary small positive constant. If δ_i equals to zero, the saturation function $sat(s(t))$ is equivalent to the sign function $sign(s(t))$. While the controlled system with direct feed-though term, the undesired chattering phenomenon affects the controlled system output directly. Hence, the saturation function should be smooth enough; in other words, the parameter δ_i should be selected properly. Therefore, the undesired chattering phenomenon can be avoided, especially if the controlled system has direct feed-though term.

According to Theorem 1, the sliding manifold is reached and substituting (19) and (20) into (11) and (12), one has

$$\dot{\eta}(t) = \overline{A}_{pidc}\eta(t) + \overline{B}_{pid1}\overline{u}(t) - \overline{B}_{pid3}\tilde{d}_g(t) - r_{pid}(t), \tag{22}$$

$$y(t) = \overline{C}_{pidc}\eta(t) + \overline{D}_{pid1}\overline{u}(t) - D\tilde{d}_g(t), \tag{23}$$

where $\overline{A}_{pidc} = \overline{A}_{pid} - \overline{B}_{pid1}K$, $\overline{C}_{pidc} = \overline{C}_{pid} - \overline{D}_{pid1}K$, $\overline{B}_{pid3} = \begin{bmatrix} 0_{n\times m} \\ D \end{bmatrix}$ and $\tilde{d}_g(t) = d_g(x,t) - \hat{d}_g(t)$.

Lemma 2. [16] *Let $(\overline{A}_{pid}, \overline{B}_{pid1})$ be the pair of the given open-loop system and $h \geq 0$ represent the prescribed degree of relative stability. The eigenvalues of the closed-loop system $\overline{A}_{pid} - \overline{B}_{pid1}(R^{-1}\overline{B}_{pid1}{}^TP)$ lie on the left of the $-h$ vertical line with the matrix P being the solution of the Riccati equation*

$$(\overline{A}_{pid} + hI_n)^T P + P(\overline{A}_{pid} + hI_n) - P\overline{B}_{pid1}R^{-1}\overline{B}_{pid1}{}^TP + Q = 0, \tag{24}$$

where the matrix I_n is an identity matrix.

In order to track the reference trajectory, the linear quadratic method is applied to the tracker design. The PI controller gain K can be described as

$$K = \begin{bmatrix} K_P & K_I \end{bmatrix} = R_c^{-1}(\overline{B}_{pid1}{}^TP + N^T),$$

where $R_c = R + \overline{D}_{pid1}{}^TQ\overline{D}_{pid1}$, $N = \overline{C}_{pid}{}^TQ\overline{D}_{pid1}$, $K_P \in \Re^{m\times n}$, and $K_I \in \Re^{m\times p}$. To design the controller gain K consisting of K_P and K_I, we temporarily do not take the perturbation estimator $\hat{d}_g(x,t)$ and the control law $\overline{u}(t)$ into consideration in (22) and (23). Both the $\hat{d}_g(x,t)$ and $\overline{u}(t)$ will be discussed based on the final-value theorem in detail.

Let the quadratic performance index for the output tracking problem be defined as

$$J = \frac{1}{2}\int_0^{t_f} \left\{[y(\tau) - r(\tau)]^T Q[y(\tau) - r(\tau)] + u^{*T}(\tau)Ru^*(\tau)\right\}d\tau, \tag{25}$$

where t_f denotes the final time, as well as $Q = 10^q I_p \in \Re^{p\times p}$ with $q \geq 0$ and $R = I_m \in \Re^{m\times m}$ are the appropriate weighting matrices. Consider the performance index in (25), to calculate the lower value for the controlled system output $y(t)$; hence, we obtain $r(t) = 0$ $(r(\tau) = 0)$ first, then utilize the final-value theorem to minimize the performance index [11]. Thus, consider Lemma 2 and (25), the algebraic Riccati equation is given by

$$\left(\overline{A}_{pid} + hI_n\right)^T P + P\left(\overline{A}_{pid} + hI_n\right) - \left(\overline{B}_{pid1}{}^T P + N^T\right)^T R^{-1}\left(\overline{B}_{pid1}{}^T P + N^T\right) + \overline{C}_{pid}{}^T Q\overline{C}_{pid} = 0. \tag{26}$$

Solving the matrix P from the algebraic Riccati equation then the control gain K can be constructed. Notice that the PI gains in K are determined based on the linear model $\left(\overline{A}_{pid}, \overline{B}_{pid1}, \overline{C}_{pid}, \overline{D}_{pid1}\right)$ first, then take the perturbation estimator $\hat{d}_g(t)$ into consideration to determine the control law $\overline{u}(t)$ in (22) and (23), based on the final-value theorem which will be discussed in detail later.

Finally, it is desirable to determine the $\overline{u}(t)$ term in (19). Since Lemma 2 is utilized, then the traditional LQAT cannot be adopted to design the control law $\overline{u}(t)$. Therefore, the final-value theorem can be utilized to find out the control law $\overline{u}(t)$. Since, the PI controller gain K has been chosen, the sliding mode is reached and $\tilde{d}(t)$ is convergence then the control law $\overline{u}(t)$ can be calculated by the final-value theorem.

Theorem 2. *The $\overline{u}(t)$ term is determined based on the integration-term-free augmented system in (22) and (23), where $\overline{u}(t) = \left[C_{pidc}(-A_{pidc})^{-1}B + D_{pid1}\right]^{\dagger}\left\{r(t) + D\hat{d}_g(t)\right\}$.*

Proof. See Appendix B. □

Finally, based on Theorem 2, the desire control law can be described as

$$u(t) = -K\eta(t) - \hat{d}_g(t) + u_{\pm}(t) + \overline{u}(t) + K_D\dot{x}\,. \tag{27}$$

Remark 3. *If the α equals to 1, the PID-type controller reduces to the PI-type controller. The control law in (27) is utilized to minimize the tracking performance in (25). Therefore, the controlled system output $y(t)$ can track the reference trajectory $r(t)$ and the tracking error can be minimized.*

3. Introduction of DNA Algorithm and Taguchi Method

3.1. DNA Algorithm

The following statements demonstrate the detailed information of DNA algorithm [31,32] operators.

A. Definition of cost function: This step defines a cost function to calculate the cost value of each individual, retain excellent chromosomes, and eliminate adverse chromosomes.

B. Reproduction: Similar to cell division, reproduction is focused on survival of the fittest. Hence, the worse chromosomes will decrease in every generation. Roulette wheel selection is one common technique to implement the proportional selection. Another way to reproduce the better population is the tournament selection. Compared with the roulette wheel selection, the tournament selection only requires the better cost values of the chromosome.

C. Crossover: After reproduction, the chromosomes mate with each other to execute the crossover operator. Crossover exchanges information between two individuals and generates two offspring. The crossover probability p_c can be decided to our demand where $p_c > 0$.

D. Mutation: In natural biological system, creatures mutate by themselves in order to adapt to the external environment. Each chromosome undergoes mutation with a fixed probability p_m where $p_m > 0$. Generally, p_m is set to be much lower than p_c in order to prevent from being unable to converge.

E. Enzyme/Virus: Enzyme and virus operators are similar to mutation operator, but the most different part is to change structure of the chromosome instead of value of the chromosome. Enzyme operator loses part of segments in chromosome; on the other hand, the virus operator increases an additional part of chromosome. Each chromosome undergoes enzyme and virus with positive probabilities p_e and p_v, respectively.

F. Termination criteria: This step provides two methods to establish a termination criterion. One is the pre-specified iteration number. Another one is reaching the tolerable error representing the algorithm that converges to the optimal solution or approaching optimal solution.

3.2. Taguchi Method

Taguchi method is a powerful and functional tool in optimization for quality [33,34,39,40]. Taguchi method uses less combination of experiments to obtain the useful information and searches the tendency of optimization to prevent from the cause of sensitive variation. The primary tools of the Taguchi method are the orthogonal array and the signal-to-noise ratio (SNR).

A. Orthogonal array: An orthogonal array can use fewer experiments than full factorial experiments. The normal expression of two-level orthogonal arrays is

$$L_{N_t}(2^{N_t-1}), \tag{28}$$

where $N_t = 2^{kt}$ denotes number of experimental runs, kt denotes a positive integer which is greater than one, 2 denotes number of levels for each factor, and $N_t - 1$ denotes number of columns in the orthogonal array.

B. SNR: Two criteria are used to determine SNR, i.e., smaller is better or larger is better. In the case of the smaller is better characteristic, let two sets of data be described by $[z_1, z_2, \ldots, z_{n_s}]$ and $[\bar{z}_1, \bar{z}_2, \ldots, \bar{z}_{n_s}]$. The mean squared deviations from the target value of the quality characteristic are described by

$$S_1 = \frac{1}{n_s}\sum_{i_s=1}^{n_s} {z_{i_s}}^2 \tag{29}$$

and

$$S_2 = \frac{1}{n_s}\sum_{i_s=1}^{n_s} {\bar{z}_{i_s}}^2. \tag{30}$$

In order to shift the mean squared deviation to a suitable situation, utilize the transformation and describe the ratio in decibels

$$\bar{S}_1 = -10\log\left(\frac{1}{n_s}\sum_{i_s=1}^{n_s} {z_{i_s}}^2\right) \tag{31}$$

and

$$\bar{S}_2 = -10\log\left(\frac{1}{n_s}\sum_{i_s=1}^{n_s} {\bar{z}_{i_s}}^2\right). \tag{32}$$

After calculating, the SNRs will be compared to decide the better level. Therefore, we can determine the better levels for each factor in less experiment. In the case of larger is better characteristic can refer to the literature [34].

4. Hybrid Taguchi and Real Coded DNA Algorithm

In this section, we are going to take advantage of DNA algorithm and Taguchi method in real coded scheme and combine with the controller design mentioned previously to select a suitable tracking controller. The detailed steps are described in Figure 1 and illustrated in the following statements.

Step 1: Coding strategy: Define a set of chromosomes including the PID gain matrices K_P, K_I, K_D in the block vector form as follows

$$C = \begin{bmatrix} K_P & K_I & K_D \end{bmatrix}. \tag{33}$$

The previously mentioned controllers can be composed of P controller, PI controller, PD controller, and PID controller. Therefore, definitions of various controller variables are $C_P^i = \begin{bmatrix} K_P^i & 0 & 0 \end{bmatrix}$, $C_{PI}^i = \begin{bmatrix} K_P^i & K_I^i & 0 \end{bmatrix}$, $C_{PD}^i = \begin{bmatrix} K_P^i & 0 & K_D^i \end{bmatrix}$, and $C_{PID}^i = \begin{bmatrix} K_P^i & K_I^i & K_D^i \end{bmatrix}$, where i denotes the i th chromosome in the whole group.

Step 2: Initialization: Before we search a solution to approximate the optimal solution, we need to generate T chromosomes for the population, which is called primitive group. To determine the different gain values in every chromosome, we select the parameters α in [0.3, 1] and q in [0, $\overline{q}$] (for example $\overline{q} = 2$) randomly to create four optimal chromosomes for each type controller, and select a gain matrix $\beta_I \in \Re^{m \times m}$ randomly to multiply the optimal chromosomes for other chromosomes until the population is reached. Each component of β_I is given a range by $[0, 1]$. Generally, the size of the primitive group depends on the problem complexity; in other words, the more complicated the problem, the larger the primitive group we need. In the experiment, we generate $T/4$ chromosomes for each type of controller.

Step 3: Reproduction: Tournament selection can be adopted to find the lower cost value for the next population.

Step 4: Crossover: The offspring chromosome has the partial characteristic from the parents after crossover. Refer to [31,34,35], a real coded crossover operator is defined and rewritten as follows

$$C_{offspring1} = \beta_c C_{parent1} + (1 - \beta_c) C_{parent2}, \tag{34}$$

where $C_{parent1}$ and $C_{parent2}$ represent different chromosomes. The parameter β_c is randomly selected and defined in a range $[0, 1]$. The crossover operator is allowed to mate with identical type controllers in the mating pool. For instance, a PI-type controller parameter C_{PI}^i only mates with the same feature chromosome.

Step 5: Choosing a proper orthogonal array: Determine the number of factors and levels to construct a suitable orthogonal array $L_4(2^3)$ for the problem demand. In the simulation, we choose three factors to make an experiment and the factors are the PID parameters. A two-level orthogonal array is studied.

Step 6: Selecting chromosomes and Taguchi experiments: This step can do ρ runs to generate ρ better chromosomes into every generation. Select a best chromosome and randomly choose another chromosome from the population. Both chromosomes are obtained to execute Taguchi method and find the better solution. In each generation, both chromosomes can be the same type of controllers or different type controllers. For example, both chromosomes $C_1(P_1, I_1, D_1)$ and $C_2(P_2, I_2, D_2)$ are the levels to be selected and each PID parameter is the factor in the orthogonal array. In this paper, the orthogonal array selects $L_4(2^3)$. The P_1, I_1 and D_1 are represented level 1 and the P_2, I_2 and D_2 are represented level 2. Calculate the SNR of each experiment in the orthogonal array, then calculate the effect of the various factors. The tracking performance is obtained and the small one is best. The formulation of SNR can be rewritten as $\rho_{\kappa j} = \frac{1}{2} \sum_{i_s=1}^{2} J_{c_{ji_s}}$ where κ represents the number of factor, j represents the number of level (J_c to be defined later), and the smaller one can be obtained. After the orthogonal array experiment, the smaller SNRs are obtained to find the best factors and the best chromosome can be found by each level. For example, level 1 is obtained in the factor P such that P_1 is selected; level 2 is obtained in the factor I, such that I_2 can be selected; level 1 is obtained in the factor D such that D_1 is selected. Based on the above description, the best chromosome is $C_{BT}(P_1, I_2, D_1)$.

Step 7: Mutation: Real coded changes its value by extending or shortening the scalar. Refer to [31,34,35], we can re-implement the mutation operator as the following

$$C_{offspring2} = C_{parent} + 2\beta_m C_{parent}, \tag{35}$$

where β_m is randomly selected in a range $[-1\,,1]$. By doing this way, it changes both the scalar and the direction to achieve mutation operator.

Step 8: Enzyme/Virus: Select two chromosomes from the population. Enzyme and virus operators can provide us with a suitable controller type. Two different type chromosomes from the pool of {P, PI, PD, PID} are randomly selected. For instance, the former operator can transform PID controller to P controller, PI controller or PD controller; the latter operator transforms P controller to PI controller, PD controller, or PID controller.

Step 9: Calculating cost value: In order to evolve the population, the cost function is employed to evaluate the value of each chromosome and the minimum one is the best chromosome. We define the cost function as

$$J_c = \int_0^{t_f} \left\{ w_1 \Big(\sum_{j_1=1}^{p} |e_{y_{j_1}}(\tau)| \Big) + w_2 \Big(\sum_{j_2=1}^{m} |u_{j_2}(\tau)| \Big) \right\} d\tau, \tag{36}$$

where $e_y(\tau) = \begin{bmatrix} e_{y_1}(\tau), & e_{y_2}(\tau), & \cdots, & e_{y_p}(\tau) \end{bmatrix}^T$ denotes the error between the output and the pre-specified trajectory, $u(\tau) = \begin{bmatrix} u_1(\tau), & u_2(\tau), & \cdots, & u_m(\tau) \end{bmatrix}^T$ denotes the control force, and J_c denotes the cost value.

Step 10: Stopping criterion: If the stopping criterion is reached, then the algorithm is terminated. Otherwise, return to Step 3 and continue to Step 10.

5. Illustrative Examples

In this section, two numerical simulations are given to illustrate the proposed fixed (optimal-based robust tracker) and flexible (HTRDNA-based robust tracker) trackers, respectively.

5.1. Fixed PID-Type Controller

To verify effectiveness of the proposed PID-based robust tracker, the following example is considered. Consider the nonlinear, Chen's chaotic system described as

$$\begin{cases} \dot{x}_1(t) = a(x_2(t) - x_1(t)) \\ \dot{x}_2(t) = (c-a)x_1(t) - x_1(t)x_3(t) + cx_2(t) + u_1(t) + d_1(x,t) \\ \dot{x}_3(t) = x_1(t)x_2(t) - bx_3(t) + u_2(t) + d_2(x,t) \end{cases}, \tag{37}$$

or in the general form

$$\dot{x}(t) = Ax(t) + B(u(t) + g(x,t) + d(x,t)), \tag{38}$$

where $A = \begin{bmatrix} -a & a & 0 \\ c-a & c & 0 \\ 0 & 0 & -b \end{bmatrix}$, $B = \begin{bmatrix} 0 & 0 \\ 1 & 0 \\ 0 & 1 \end{bmatrix}$, $x(t) = \begin{bmatrix} x_1(t) \\ x_2(t) \\ x_3(t) \end{bmatrix}$, $u^*(t) = \begin{bmatrix} u_1{}^*(t) \\ u_2{}^*(t) \end{bmatrix}$, $g(x,t) = \begin{bmatrix} -x_1(t)x_3(t) \\ x_1(t)x_2(t) \end{bmatrix}$, $d(x,t) = \begin{bmatrix} d_1(x,t) \\ d_2(x,t) \end{bmatrix}$, in which $a = 35$, $b = 3$, $c = 28$, $x \in \Re^3$, $u(t) \in \Re^2$ and the initial condition is selected as $x(0) = \begin{bmatrix} -0.5 & 0.2 & 0.3 \end{bmatrix}^T$. The bounded nonlinear perturbation and the reference trajectory $r(t)$ are, respectively, given by

$$d_g(x,t) = \begin{bmatrix} \cos(x_1) & 0 \\ 0 & \sin(x_2) \end{bmatrix} \begin{bmatrix} 0.3 & 0 & 0 \\ 0 & -0.4 & 0.1 \end{bmatrix} x(t) + g(x,t)$$

and

$$r(t) = \begin{cases} \begin{bmatrix} 5\sin(2\pi t/1.5) & 5\sin(2\pi t/1.5) \end{bmatrix}^T, & t \le 1.5\,\text{sec} \\ \begin{bmatrix} 5 & 5 \end{bmatrix}^T, & t > 1.5\,\text{sec} \end{cases}.$$

Let the output be represented by the general form

$$y(t) = Cx(t) + Du(t), \tag{39}$$

where $C = \begin{bmatrix} -0.5 & 5 & 0 \\ 0 & 0 & 0.5 \end{bmatrix}$, $D = \begin{bmatrix} 0.1 & 0 \\ 0 & 0.2 \end{bmatrix}$, $y(t) = \begin{bmatrix} y_1(t) \\ y_2(t) \end{bmatrix}$, in which $y \in \Re^2$.

We set the matrix pair $\{Q, R\} = \{10^3 I_2, I_2\}$ for the controller design, $k_o = 350$, $h = 5$ and $\alpha = 0.8$ to yield $K_D = \begin{bmatrix} 0 & 0.2 & 0 \\ 0 & 0 & 0.2 \end{bmatrix}$, $M = I_3 - BK_D = \begin{bmatrix} 1 & 0 & 0 \\ 0 & 0.8 & 0 \\ 0 & 0 & 0.8 \end{bmatrix}$, $\overline{A}_{pid} = \begin{bmatrix} -35 & 35 & 0 & 0 & 0 \\ -8.75 & 35 & 0 & 0 & 0 \\ 0 & 0 & -3.75 & 0 & 0 \\ -0.675 & 5.7 & 0 & 0 & 0 \\ 0 & 0 & 0.35 & 0 & 0 \end{bmatrix}$, $\overline{B}_{pid} = \begin{bmatrix} 0 & 0 \\ 1.25 & 0 \\ 0 & 1.25 \\ 0.125 & 0 \\ 0 & 0.25 \end{bmatrix}$, $\overline{C}_{pid} = \begin{bmatrix} -0.675 & 5.70 & 0 & 0 & 0 \\ 0 & 0 & 0.35 & 0 & 0 \end{bmatrix}$, $D_{pid1} = \begin{bmatrix} 0.125 & 0 \\ 0 & 0.25 \end{bmatrix}$, $D_{pid2} = \begin{bmatrix} 0.025 & 0 \\ 0 & 0.05 \end{bmatrix}$, $C_s = \begin{bmatrix} 0 & 0.8 & 0 & 0 & 0 \\ 0 & 0 & 0.8 & 0 & 0 \end{bmatrix}$, $\gamma = 100$, $\sigma = 0.1$ and $\delta = 10^{-3}$.
The PI gain matrices can be obtained as below

$$K = \begin{bmatrix} K_P & K_I \end{bmatrix} = \begin{bmatrix} -5.503 & 46.1477 & 0 & 82.7264 & 0 \\ 0 & 0 & 1.3867 & 0 & 40.1473 \end{bmatrix},$$

where $K_P \in \Re^{2\times3}$ and $K_I \in \Re^{2\times2}$. The sliding surface and fixed PID-type controller are given in (13) and (27), respectively.

Figures 2–4 demonstrate the tracking performance between the controlled system output $y(t)$ and the pre-specify trajectory $r(t)$. The sliding surface is shown in Figure 5. The estimation error between perturbation estimator and perturbation is shown in Figure 6. Figures 2–6 demonstrate a satisfied performance based on the proposed robust tracker for the system with unknown perturbation.

5.2. Flexible PID-Type Controller Based on the HTRDNA

To improve the tracking performance of the proposed PID-based robust tracker, the proposed HTRDNA is adopted. Consider the same Chen's chaotic system given in Section 5.1. For searching the best cost value during the iterative process, we define the cost function as

$$J_c = \int_0^{t_f} \left\{ w_1\left(\sum_{j_1=1}^{p} |e_{y_{j_1}}(\tau)|\right) + w_2\left(\sum_{j_2=1}^{m} |u_{j_2}(\tau)|\right) \right\} d\tau, \tag{40}$$

where $e_y(\tau) = \begin{bmatrix} e_{y_1}(\tau), & e_{y_2}(\tau), & \cdots, & e_{y_p}(\tau) \end{bmatrix}^T$ denotes the error between the output and the pre-specified trajectory, $u(\tau) = \begin{bmatrix} u_1(\tau), & u_2(\tau), & \cdots, & u_m(\tau) \end{bmatrix}^T$ denotes the control force, J_c denotes the cost value.

Here, we hope to apply the HTRDNA algorithm to seek for the best one from four kinds of PID-type controllers. The parameters are chosen as follows: The maximum iteration number is 100, probability of crossover $p_c = 0.5$, probability of mutation $p_m = 0.01$, probability of enzyme $p_e = 0.01$, probability of virus $p_v = 0.01$, the orthogonal array select $L_4(2^3)$, the weighting $w_1 = 1$ and $w_2 = 10^{-3}$. The resultant controller selected based on the HTRDNA algorithm is the PID-type controller and its parameters are

$K_P = \begin{bmatrix} -5.0668 & 49.8911 & 0 \\ 0 & 0 & 2.4501 \end{bmatrix}$, $K_I = \begin{bmatrix} 102.4782 & 0.0001 \\ 0.0002 & 50.7663 \end{bmatrix}$ and $K_D = \begin{bmatrix} 0 & 0.0081 & 0.0061 \\ 0 & 0.0019 & 0.0096 \end{bmatrix}$.
The sliding surface and fixed PID-type controller are given in (13) and (27), respectively.

Figures 7–11 demonstrate a quite satisfied tracking performance based on the proposed method. According to Figures 3 and 8, the proposed HTRDNA algorithm can improve the error performance by considering the performance index in (40). Figures 4 and 9 show the control input without undesired chartering phenomenon by using the proposed control law (27). Compare Figure 4 with Figure 9, Figure 9 shows that the control input is constrained by the performance index in (40). Figures 6 and 11 show that the error of perturbation estimation is converged. The simulation results demonstrate the validity of the proposed perturbation estimator method. Furthermore, based on the cost function (40), Figure 12 shows that the proposed flexible PID-type controller outperforms the fixed PID-type controller. In addition, Figure 12 shows that the proposed HTRDNA algorithm outperforms the real code DNA (RDNA) algorithm. Consider the performance index (40) to Section 5.1, the cost value is 0.2129. After HTRDNA algorithm optimization, the cost value is 0.1793. Compare Section 5.1 with Section 5.2, the proposed HTRDNA algorithm can optimize the controller and improve the tracking performance. Based on the above description, the newly proposed HTRDNA algorithm can improve the performance for the proposed controller.

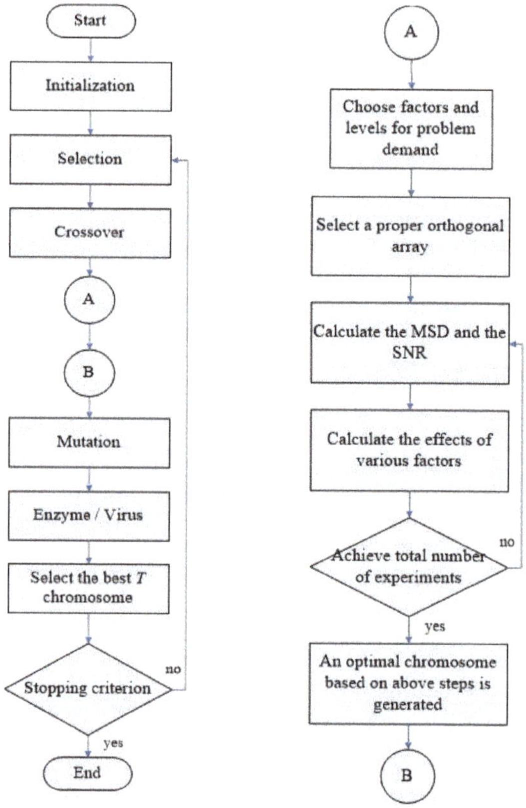

Figure 1. Flow chart for HTRDNA algorithm.

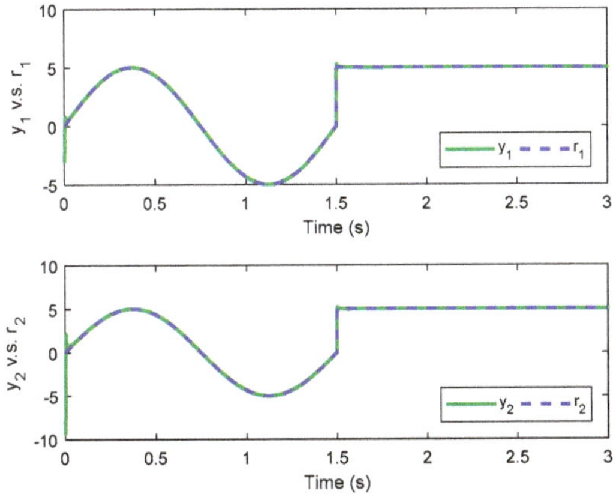

Figure 2. Time responses of the closed-loop system with the fixed PID controller and unknown perturbation.

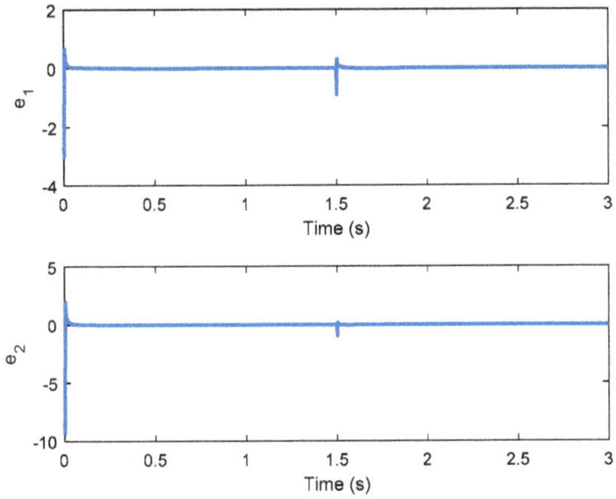

Figure 3. Tracking errors of the closed-loop system with the fixed PID controller and unknown perturbation.

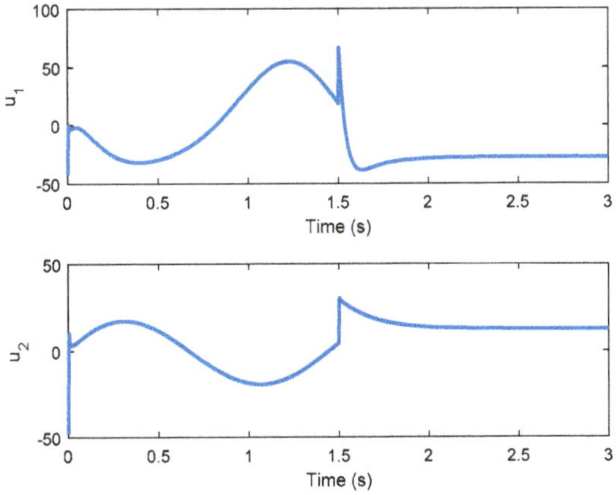

Figure 4. Control inputs based on the fixed PID controller and unknown perturbation.

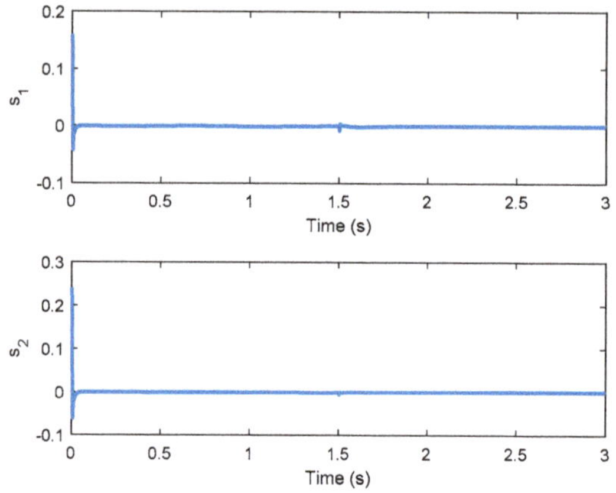

Figure 5. Sliding manifolds for the fixed PID controller with unknown perturbation.

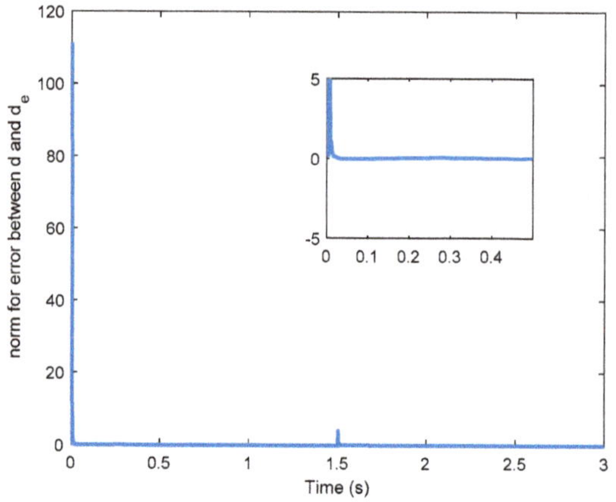

Figure 6. Error between unknown and estimated perturbations.

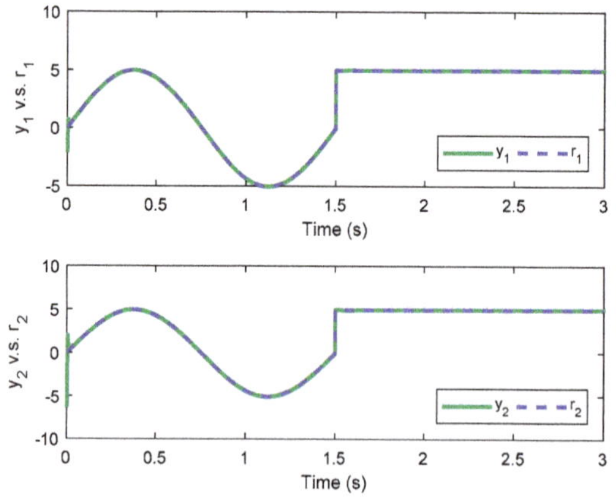

Figure 7. Time responses of the closed-loop system with the flexible PID controller and unknown perturbation.

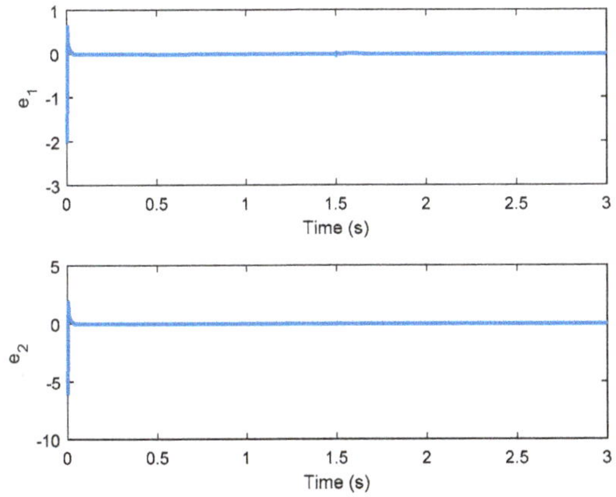

Figure 8. Tracking errors of the closed-loop system with the flexible PID controller and unknown perturbation.

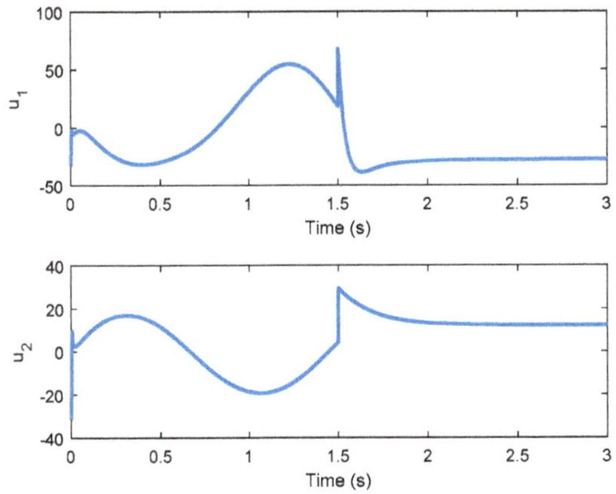

Figure 9. Control inputs based on the flexible PID controller and unknown perturbation.

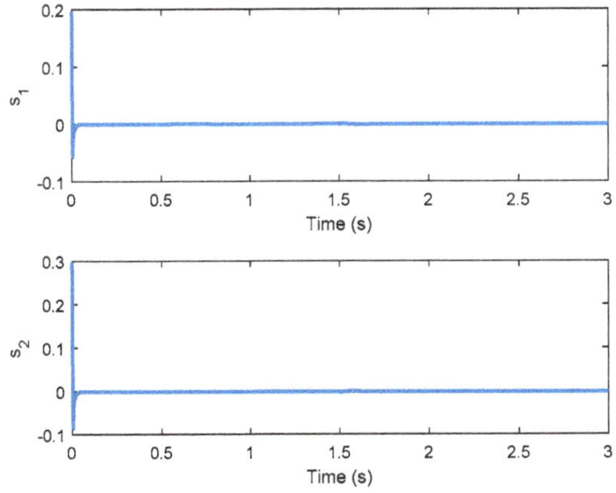

Figure 10. Sliding manifolds for the flexible PID controller with unknown perturbation.

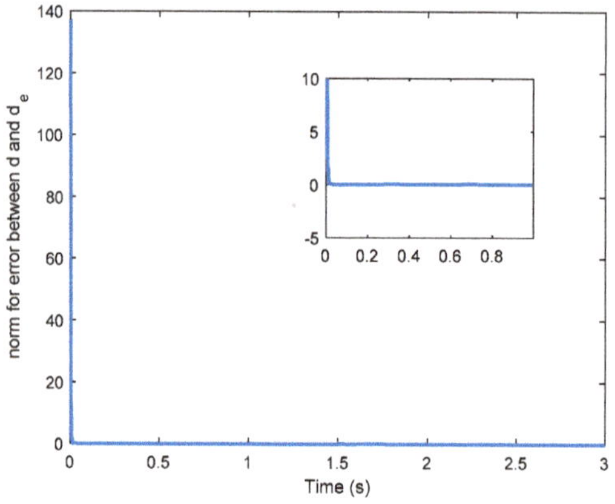

Figure 11. Error between unknown and estimated perturbations.

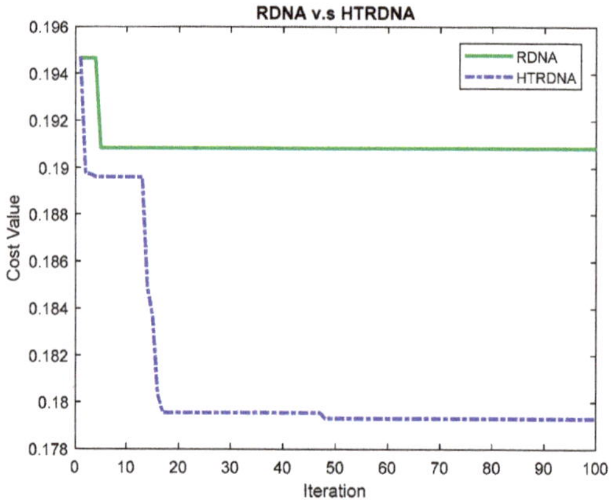

Figure 12. Evolution of RDNA and HTRDNA algorithm.

6. Conclusions

A robust tracker design for a class of nonlinear controlled systems with/without direct feed-through term and unknown nonlinear perturbation is proposed in this paper. Based on LQAT, by taking linear transformation and augmented state, a simple approach for the PID-type controller with SMC and perturbation estimator is proposed. The designed perturbation estimator is employed to eliminate the unknown nonlinear perturbation so that the better performance can be achieved. To improve the efficiency of real coded DNA algorithm, this paper utilizes the advantage of the Taguchi method to real coded DNA algorithm so that the HTRDNA algorithm is newly proposed for the PID controller optimization. Due to the SMC with fast response, SMC is employed to cope with the nonlinear perturbation and then HTRDNA algorithm can be utilized to tune the PID controller type and its parameters. Simulation results demonstrate the validity of our proposed method.

Author Contributions: All authors contributed to the paper. J.-S.F. wrote the manuscript with the supervision from J.-J.Y. and J.S.-H.T. C.-H.T. and S.-M.G. are responsible for the simulation of the proposed robust tracker.

Funding: This work was financially supported by the Ministry of Science and Technology, Taiwan, under MOST-107-2221-E-167 -032 -MY2.

Conflicts of Interest: The authors declare no conflict of interest.

Appendix A

Proof of Theorem A1. Substitute (19) and (20) into the derivative of sliding surface in (13) to obtain

$$\dot{s}(t) = \tilde{d}_g(t) - \gamma s(t). \tag{A1}$$

Differentiating (17), one has

$$\begin{aligned}\dot{\hat{d}}(t) &= k_o\big(\dot{s}(t) + \gamma s(t)\big) = k_o\big(\tilde{d}_g(t) - \gamma s(t) + \gamma s(t)\big)\\ &= k_o\tilde{d}_g(t).\end{aligned} \tag{A2}$$

Substituting (16) and (A2) into the differentiation of (20) yields

$$\begin{aligned}\dot{\tilde{d}}_g(t) &= \dot{d}_g(x,t) - \dot{\hat{d}}_g(t) = \dot{d}_g(x,t) - k_o\tilde{d}_g(t)\\ &= -k_o\tilde{d}_g(t).\end{aligned} \tag{A3}$$

If the gain k_o is selected to be a positive value, the error of (20) can converge and approximate to zero. In other words, the estimated perturbation can approximate to the unknown perturbation at the steady state.

Consider a candidate Lyapunov function as

$$v(s) = \frac{1}{2}s^T s, \tag{A4}$$

and taking the derivative of $v(s)$ in (A4) gives

$$\begin{aligned}\dot{v}(s) &= s^T\dot{s} = s^T\big(\tilde{d}_g(x,t) - \gamma s - \sigma sat(s(t))\big)\\ &\le \|\tilde{d}_g(x,t)\|\|s\| - \gamma\|s\|^2 - \sigma\|s\|\\ &\le -\gamma\|s\|^2 - \sigma\|s\|.\end{aligned} \tag{A5}$$

Equations (A3)–(A5) show that the sliding mode states can reach the defined sliding manifold in finite time with the given parameters $\gamma > 0$ and $\sigma > 0$; therefore, (17) can estimate the unknown external perturbation and eliminate its impact directly. In addition, when $\tilde{d}_g(t)$ equals or closes to zero, the controller in (19) can achieve a desired tracking performance. □

Appendix B

Proof of Theorem A2. Consider a linear time-invariant system with the PI-type controller and underdetermined $\overline{u}(t)$ term described by

$$\dot{x}(t) = Ax(t) + B\left(\overline{u}(t) - K_P x(t) - K_I\int e_y(t)dt\right), \tag{A6}$$

$$y(t) = Cx(t) + D\left(\overline{u}(t) - K_P x(t) - K_I\int e_y(t)dt\right). \tag{A7}$$

Take the Laplace transform of the tracking error to obtain the following equations

$$E_y(s) = Y(s) - R_s = \left\{(C - DK_P)[sI_n - (A - BK_P)]^{-1}B + D\right\}\left(\frac{\overline{U}_s}{s} - K_I\frac{E(s)}{s}\right) - \frac{R_s}{s}, \tag{A8}$$

where $\overline{U}_s$ and R_s are the steady-state values of $\overline{u}(t)$ and $r(t)$, respectively, during any time period, if $\overline{u}(t)$ and $r(t)$ change slightly relative to the high gain property controlled system dynamics. Using the final-value theorem to (A8), one has

$$\lim_{s\to 0} sE_y(s) = \lim_{s\to 0} s\left[W\left(\frac{\overline{U}_s}{s} - K_i\frac{E(s)}{s}\right) - \frac{R_s}{s}\right] = \lim_{s\to 0}\left[W\left(\overline{U}(s) - K_iE(s)\right) - R_s\right], \tag{A9}$$

where

$$W = (C - DK_P)[sI_n - (A - BK_P)]^{-1}B + D. \tag{A10}$$

Rearrange (A9) to have

$$\lim_{s\to 0}(sI_n + K_IW)E_y(s) = \lim_{s\to 0}\left(W\overline{U}_s - R_s\right),$$

which implies

$$\lim_{s\to 0}\left\{(C - DK_P)[sI_n - (A - BK_P)]^{-1}B + D\right\}\overline{U}_s - R_s = 0$$

for $\lim_{s\to 0} sE_y(s) = 0$. From (A10), we can infer that it is sufficient to derive the controller $\overline{u}(t)$ in (22) and (23) by applying the final-value theorem without the integral term.

According to Theorem 1 and Theorem 2, SMC is reached and the perturbation is estimated by the perturbation estimator. Then, take Laplace transforms of (22) and (23) without integral term to obtain

$$\begin{aligned} Y(s) &= C_{pidc}(sI_n - A_{pidc})^{-1}B_{pid}\frac{\overline{U}_s}{s} + D_{pid1}\left(\frac{\overline{U}_s}{s} - \frac{\hat{D}_{gs}}{s}\right) \\ &= \left[C_{pidc}(sI_n - A_{pidc})^{-1}B_{pid} + D_{pid1}\right]\frac{\overline{U}_s}{s} - D\frac{\hat{D}_{gs}}{s}, \end{aligned} \tag{A11}$$

where $\hat{D}_{g_s}$ is the steady-state values of $\hat{d}_g(t)$, during any time period, if $\hat{d}_g(t)$ changes slightly relative to the high gain property controlled system dynamics. Applying the final-value theorem to the tracking error and forcing it to be zero yields

$$\begin{aligned} \lim_{s\to 0} sE_y(s) &= \lim_{s\to 0} s(Y(s) - R_s) \\ &= \left[C_{pidc}(-A_{pidc})^{-1}B_{pid} + D_{pid1}\right]\overline{U}_s - D\hat{D}_{g_s} - R_s \\ &= 0, \end{aligned}$$

so that in general, one has

$$\overline{u}(t) = \left[C_{pidc}(-A_{pidc})^{-1}B + D_{pid1}\right]^{\dagger}\left\{r(t) + D\hat{d}_g(t)\right\}, \tag{A12}$$

where

$$A_{pidc} = A_{pid} - B_{pid}K_P,$$

□

References

1. Ang, K.; Chong, G.; Li, Y. PID control system analysis, design, and technology. *IEEE Trans. Control Syst. Technol.* **2005**, *13*, 559–576.
2. Åström, K.J.; Hägglund, T. *PID Controllers: Theory, Design, and Tuning*; ISA, The Instrumentation, Systems, and Automation Society: Pittsburgh, PA, USA, 1995.
3. Gaing, Z.L. A particle swarm optimization approach for optimum design of PID controller in AVR system. *IEEE Trans. Energy Convers.* **2004**, *19*, 384–391. [CrossRef]
4. Nazaruddin, Y.Y.; Andrini, A.D.; Anditio, B. PSO Based PID Controller for Quadrotor with Virtual Sensor. *IFAC-PapersOnLine* **2018**, *51*, 358–363. [CrossRef]

5. Abdelaziz, T.H.S.; Valasek, M. Pole-placement for SISO Linear Systems by State-derivative Feedback. *IEE Proc. Control Theory Appl.* **2004**, *151*, 377–385. [CrossRef]
6. Cardim, R.; Teixeira, M.C.M.; Assuncao, E.; Covacic, M.R. Design of state-derivative feedback controllers using a state feedback control design. *IFAC Proc. Vol.* **2007**, *40*, 22–27. [CrossRef]
7. Zheng, F.; Wang, Q.G.; Lee, T.H. On the design of multivariable PID controllers via LMI approach. *Automaticca* **2002**, *38*, 517–526. [CrossRef]
8. Rosinová, D.; Veselý, V. Robust pid decentralized controller design using lmi. *IFAC Proc. Vol.* **2006**, *39*, 53–58. [CrossRef]
9. Wang, J.; Zhang, Q.; Xiao, D. PD feedback H_∞ control for uncertain singular neutral systems. *Adv. Differ. Equ.* **2016**, *2016*, 29. [CrossRef]
10. Shariati, A.; Taghirad, H.D.; Fatehi, A. A neutral system approach to H_∞ PD/PI controller design of processes with uncertain input delay. *J. Process Control* **2014**, *24*, 144–157. [CrossRef]
11. Singla, M.; Shieh, L.S.; Song, G.; Xie, L.; Zhang, Y. A new optimal sliding mode controller design using scalar sign function. *ISA Trans.* **2014**, *53*, 267–279. [CrossRef]
12. Madsen, J.M.; Shieh, L.S.; Guo, S.M. State-Space Digital PID Controller Design for Multivariable Analog Systems with Multiple Time Delays. *Asian J. Control* **2006**, *8*, 161–173. [CrossRef]
13. Ebrahimzadeh, F.; Tsai, J.S.H.; Liao, Y.T.; Chung, M.C.; Guo, S.M.; Shieh, L.S.; Wang, L. A generalised optimal linear quadratic tracker with universal applications—Part 1: Continuous-time systems. *Int. J. Syst. Sci.* **2017**, *48*, 376–396. [CrossRef]
14. Tsai, J.S.H.; Wang, H.H.; Guo, S.M.; Shieh, L.S.; Canelon, J.I. A case study on the universal compensation improvement mechanism: A robust PID filter-shaped optimal PI tracker for systems with/without disturbances. *J. Frankl. Inst.* **2018**, *355*, 3583–3618. [CrossRef]
15. Chen, W.H.; Ballance, D.J.; Gawthrop, P.J.; O'Reilly, J. A nonlinear disturbance observer for robotic manipulators. *IEEE Trans. Ind. Electron.* **2000**, *47*, 932–938. [CrossRef]
16. Shieh, S.L.; Dib, M.H.; Sekar, G. Continuous-time quadratic regulators and pseudo-continuous-time quadratic regulators with pole placement in a specific region. *IEE Proc. D-Control Theory Appl.* **1987**, *134*, 338–346. [CrossRef]
17. Chang, W.D.; Yan, J.J. Adaptive robust PID controller design based on a sliding mode for uncertain chaotic systems. *Chaos Solitons Fractals* **2005**, *26*, 167–175. [CrossRef]
18. Fnaiech, M.A.; Betin, F.; Capolino, G.A.; Fnaiech, F. Fuzzy logic and sliding-mode controls applied to six-phase induction machine with open phases. *IEEE Trans. Ind. Electron.* **2010**, *57*, 354–364. [CrossRef]
19. Mondal, S.; Mahanta, C. Chattering free adaptive multivariable sliding mode controller for systems with matched and mismatched uncertainty. *ISA Trans.* **2013**, *52*, 335–341. [CrossRef]
20. Singh, Y.; Santhakumar, M. Inverse dynamics and robust sliding mode control of a planar parallel (2-PRP and 1-PPR) robot augmented with a nonlinear disturbance observer. *Mech. Mach. Theory* **2015**, *92*, 29–50. [CrossRef]
21. Li, D. Nonlinear disturbance observer-based robust control for spacecraft formation flying. *Aerosp. Sci. Technol.* **2018**, *76*, 82–90.
22. Huang, J.; Ri, S.; Fukuda, T.; Wang, Y. A Disturbance Observer based Sliding Mode Control for a Class of Underactuated Robotic System with Mismatched Uncertainties. *IEEE Trans. Autom. Control* **2019**, *64*, 2480–2487. [CrossRef]
23. Ahmed, N.; Chen, M. Sliding mode control for quadrotor with disturbance observer. *Adv. Mech. Eng.* **2018**, *10*. [CrossRef]
24. Qian, R.; Luo, M.; Zhao, Y.; Zhao, J. Novel Adaptive Sliding Mode Control with Nonlinear Disturbance Observer for SMT Assembly Machine. *Math. Probl. Eng.* **2016**, *2016*, 14. [CrossRef]
25. Wang, J.; Li, S.; Yang, J.; Wu, B.; Li, Q. Extended state observer-based sliding mode control for PWM-based DC-DC buck power converter systems with mismatched disturbances. *IET Control Theory Appl.* **2015**, *9*, 579–586. [CrossRef]
26. Li, Y.; Zheng, Q.; Yang, L. Design of robust sliding mode control with disturbance observer for multi-axis coordinated traveling system. *Comput. Math. Appl.* **2012**, *64*, 759–765. [CrossRef]
27. Hast, M.; Åström, K.J.; Bernhardsson, B.; Boyd, S. PID design by convex-concave optimization. In Proceedings of the 2013 European Control Conference (ECC), Zurich, Switzerland, 17–19 July 2013; pp. 4460–4465.

28. Mercader, P.; Åström, K.J.; Banos, A.; Hägglund, T. Robust PID design based on QFT and convex–concave optimization. *IEEE Trans. Control Syst. Technol.* **2016**, *25*, 441–452. [CrossRef]
29. Karimi, A.; Kammer, C. A data-driven approach to robust control of multivariable systems by convex optimization. *Automatica* **2017**, *85*, 227–233. [CrossRef]
30. Vijay, M.; Jena, D. PSO based neuro fuzzy sliding mode control for a robot manipulator. *J. Electr. Syst. Inf. Technol.* **2016**, *4*, 243–256. [CrossRef]
31. Wu, C.T.; Tien, J.P.; Li, T.H.S. Integration of DNA and real coded GA for the design of PID-like fuzzy controllers. In Proceedings of the 2012 IEEE International Conference on Systems, Man, and Cybernetics (SMC), Seoul, Korea, 14–17 October 2012; pp. 2809–2814.
32. Yourui, H.; Chen, X.; Yihua, H. Optimization for parameter of PID based on DNA genetic algorithm. In Proceedings of the 2005 International Conference on Neural Networks and Brain, Beijing, China, 13–15 October 2005; pp. 859–861.
33. Chou, J.H.; Chen, S.H.; Li, J.J. Application of the Taguchi-genetic method to design an optimal grey-fuzzy controller of a constant turning force system. *J. Mater. Process. Technol.* **2000**, *105*, 333–343. [CrossRef]
34. Tsai, J.T.; Liu, T.K.; Chou, J.H. Hybrid Taguchi-genetic algorithm for global numerical optimization. *IEEE Trans. Evol. Comput.* **2004**, *8*, 365–377. [CrossRef]
35. Chen, J.L.; Chang, W.D. Feedback linearization control of a two-link robot using a multi-crossover genetic algorithm. *Expert Syst. Appl.* **2009**, *36*, 4154–4159. [CrossRef]
36. Zhou, X.; Gao, H.; Zhao, B.; Zhao, L. A GA-based parameters tuning method for an ADRC controller of ISP for aerial remote sensing applications. *ISA Trans.* **2018**, *81*, 318–328. [CrossRef]
37. Feng, H.; Yin, C.B.; Weng, W.W.; Ma, W.; Zhou, J.J.; Jia, W.H.; Zhang, Z.L. Robotic excavator trajectory control using an improved GA based PID controller. *Mech. Syst. Signal Process* **2018**, *105*, 153–168. [CrossRef]
38. Zemmit, A.; Messalti, S.; Harrag, A. A new improved DTC of doubly fed induction machine using GA-based PI controller. *Ain Shams Eng. J.* **2017**, *9*, 1877–1885. [CrossRef]
39. Phadke, M.S. *Quality Engineering Using Robust Design*; Prentice Hall PTR: Upper Saddle River, NJ, USA, 1995.
40. Ross, P.J. *Taguchi Techniques for Quality Engineering: Loss Function, Orthogonal Experiments*; Parameter and Tolerance Design, McGraw Hill: New York, NY, USA, 1988.
41. Lin, J.S.; Huang, C.F.; Liao, T.L.; Yan, J.J. Design and implementation of digital secure communication based on synchronized chaotic systems. *Digit. Signal Process.* **2010**, *20*, 229–237. [CrossRef]

mathematics

MDPI

Article

Analysis of a SEIR-KS Mathematical Model For Computer Virus Propagation in a Periodic Environment

Aníbal Coronel [1,*], Fernando Huancas [2], Ian Hess [1], Esperanza Lozada [1] and Francisco Novoa-Muñoz [3]

[1] Departamento de Ciencias Básicas, Facultad de Ciencias, Universidad del Bío-Bío, Campus Fernando May, 3780000 Chillán, Chile; ihess@egresados.ubiobio.cl (I.H.); elozada@udec.cl (E.L.)
[2] Departamento de Matemática, Facultad de Ciencias Naturales, Matemática y del Medio Ambiente, Universidad Tecnológica Metropolitana, Las Palmeras 3360, 8330378 Ñuñoa-Santiago, Chile; fhuancas@utem.cl
[3] Departamento de Estadística, Facultad de Ciencias, Universidad del Bío-Bío, 4051381 Concepción, Chile; fnovoa@ubiobio.cl
* Correspondence: acoronel@ubiobio.cl; Tel.: +56-42-2463259

Received: 16 April 2020; Accepted: 7 May 2020; Published: 11 May 2020

Abstract: In this work we develop a study of positive periodic solutions for a mathematical model of the dynamics of computer virus propagation. We propose a generalized compartment model of SEIR-KS type, since we consider that the population is partitioned in five classes: susceptible (S); exposed (E); infected (I); recovered (R); and kill signals (K), and assume that the rates of virus propagation are time dependent functions. Then, we introduce a sufficient condition for the existence of positive periodic solutions of the generalized SEIR-KS model. The proof of the main results are based on a priori estimates of the SEIR-KS system solutions and the application of coincidence degree theory. Moreover, we present an example of a generalized system satisfying the sufficient condition.

Keywords: periodic solutions; positive solutions; SEIR-KS model; computer virus model

MSC: 34K13; 34D23

1. Introduction

1.1. Scope

In the last decades, due to its theoretical and practical importance and significance, the mathematical models for dynamics of propagation for epidemics have been extensively studied, see for instance [1–10] and references in those works. In particular, mathematical models are powerful tools since it permits to explain, estimate and simulate the spread of infectious disease propagation, and consequently help to design and test control strategies like an optimal time of vaccination.

From the historical point of view, the earliest mathematical models in epidemiology were introduced in 1927 [11]. Following the presentation given in [12], we have that the basic idea considered in [11], in order to describe the dynamics of a virus, was the partition of the total population N in three classes: the susceptible class S formed for those individuals capable of contracting the disease and becoming themselves infectives; the infective class I formed for those individuals capable of transmitting the disease to susceptibles; the removed or recovered class R formed for those individuals which having contracted the disease, have died or, are permanently immune, or have been isolated, thus being unable to further transmit the disease. Moreover, they consider the three assumptions:

the period of the epidemic is too short such that N is constant, the transfer process from S to I is modeled by the mass action law and the transfer process from I to R is of exponential decay type. Then, the called SIR model is given by the following system

$$\frac{dS(t)}{dt} = -kI(t)S(t), \qquad \frac{dI(t)}{dt} = kI(t)S(t) - \lambda I(t), \qquad \frac{dR(t)}{dt} = \lambda I(t),$$

where k and λ are some positive constants. A particular case of SIR model is the well known SIS model, which is deduced by considering the partition of the population in two classes of individuals: susceptible and infected. Afterwards, numerous generalizations are given by several authors, who have improved the SIR mathematical model by incorporating for instance the vital dynamics, a generalized transmission forces, other classes of individuals and vaccination.

It is well known that the outbreaks of parasite population, which generate the epidemics occur around the same time of each year. Then becomes natural to study the periodicity or model these diseases by incorporating periodic functions into the epidemic models. For instance, in the case of the SIR model the periodic models are introduced by considering the facts that k and λ are time dependent periodic functions.

On the other hand, the compartmental models were introduced for biological epidemics. However, by the newest observation that the diffusion of biological virus is analogous to several processes in other areas, the ideas have been widely adapted and used to describe other phenomenon. For instance, the computer virus propagation in a network [13–18]. In particular, in this paper our aim is to study the periodicity of the mathematical model for virus propagation introduced in [18].

1.2. The Generalized SEIR-KS Mathematical Model

In [18] the authors construct a compartmental model for computer virus propagation. They consider that the population of individuals is given by computers or nodes in a network which are in corresponding communications all the time. The population is partitioned in five classes: the susceptible class S formed by the nodes which are virus-free uninfected; the exposed class E formed by the nodes which are infected, but the virus is latent; the infected class I formed by the nodes which are infected and the virus is breaking out; the recovered class R formed by the nodes which have recovered from virus infection and acquired immunization; and the kill signals class K formed by special nodes, which are a sort of anti-virus epidemic riding on the back of the virus propagation and all of them constitute a new compartment, which is generated among the infectious nodes then they can spontaneously transmit it to their neighboring nodes. The dynamic of computer virus transmission is studied by considering the following list of assumptions:

- (A1) The network at time t is formed by a total of $N(t)$ nodes. Then, we have the following relation $N(t) = S(t) + E(t) + I(t) + R(t) + K(t)$ at each time t.
- (A2) There is a behavior similar to vital dynamics of biological virus. More specifically, related with births and deaths, there is two characteristics in the process: (i) the new nodes are connected to the network at constant rate b and a fraction p are of susceptible type and the remaining fraction $q = 1 - p$ are of exposed type; and (ii) each node, by system crash or network interruption, are disconnected from the network at constant rate μ.
- (A3) The dynamics of exposed nodes are characterized by three facts: (i) the susceptible nodes are transformed in exposed nodes with probability per unit time $\beta E(t)$ with β a constant; (ii) the exposed nodes are converted into infected ones at constant rate α; and (iii) the exposed nodes are converted into kill signals ones at constant rate χ.
- (A4) The infected nodes are converted into kill signals nodes or recovered ones at constant rates γ and ε, respectively.
- (A5) The kill signal nodes satisfy two additional premises: (i) the susceptible nodes receive the kill signal and converted into recovered ones with probability $\phi K(t)$; and (ii) the infected nodes receives and relays the kill signal nodes with probability $\delta K(t)$. Here ϕ and δ are constants.

Then, the following ordinary differential equation system

$$\frac{dS(t)}{dt} = pb - \beta S(t)E(t) - \phi S(t)K(t) - \mu S(t), \tag{1a}$$
$$\frac{dE(t)}{dt} = qb + \beta S(t)E(t) - \alpha E(t) - \chi E(t) - \mu E(t), \tag{1b}$$
$$\frac{dI(t)}{dt} = \alpha E(t) - \delta I(t)K(t) - \gamma I(t) - \varepsilon I(t) - \mu I(t), \tag{1c}$$
$$\frac{dK(t)}{dt} = \delta I(t)K(t) + \gamma I(t) + \chi E(t) - \mu K(t), \tag{1d}$$
$$\frac{dR(t)}{dt} = \phi S(t)K(t) + \varepsilon I(t) - \mu R(t), \tag{1e}$$

is introduced as the mathematical model for computer virus propagation.

In this work, with the purpose to study the existence of periodic solutions for systems of Equation (1), we consider a more general model by assuming that constants on the assumptions (A2)–(A5) are time dependent real functions, i.e., the parameters b, p, q, p, α, β, γ, χ, ϕ, δ, μ and ε are time dependent real functions. More precisely, we are motivated by the analysis of the following generalized model:

$$\frac{dS(t)}{dt} = p(t)b(t) - \beta(t)S(t)E(t) - \phi(t)S(t)K(t) - \mu(t)S(t), \tag{2a}$$
$$\frac{dE(t)}{dt} = q(t)b(t) + \beta(t)S(t)E(t) - \alpha(t)E(t) - \chi(t)E(t) - \mu(t)E(t), \tag{2b}$$
$$\frac{dI(t)}{dt} = \alpha(t)E(t) - \delta(t)I(t)K(t) - \gamma(t)I(t) - \varepsilon(t)I(t) - \mu(t)I(t), \tag{2c}$$
$$\frac{dK(t)}{dt} = \delta(t)I(t)K(t) + \gamma(t)I(t) + \chi(t)E(t) - \mu(t)K(t), \tag{2d}$$
$$\frac{dR(t)}{dt} = \phi(t)S(t)K(t) + \varepsilon(t)I(t) - \mu(t)R(t). \tag{2e}$$

We observe that the system in Equation (2) can be uncoupled in the study of the system in Equation (2)a–e. Indeed, it is the strategy considered in [18] to analyze the stability. However, to study the existence of periodic solutions is more convenient to consider the full system, since it is not straightforward the fact that the existence of positive periodic solutions for Equation (2)a–d implies the existence of positive periodic solution for Equation (2)e.

1.3. Reformulation of System in Equation (2) as Operator Equation

Firstly, we introduce a change of variable such that the system in Equation (2) is replaced by an equivalent system. Then, we reformulate the new system as seen in Equation (4) as an operator equation which will be analyzed by the topological degree theory.

For S, E, I, K and R satisfying the system in Equation (2), we consider the new functions S^*, E^*, I^*, K^* and R^* defined explicitly by the relation

$$(S, E, I, K, R)(t) = \Big(\exp(S^*(t)), \exp(E^*(t)), \exp(I^*(t)), \exp(K^*(t)), \exp(R^*(t))\Big). \tag{3}$$

Then, by differentiation in Equation (3) and using the fact that (S, E, I, K, R) satisfy the mathematical model in Equation (2), we deduce that $(S^*, E^*, I^*, K^*, R^*)$ is a solution of the system

$$\frac{dS^*(t)}{dt} = p(t)b(t)\exp(-S^*(t)) - \beta(t)\exp(E^*(t)) - \phi(t)\exp((K^* - S^*)(t)) - \mu(t), \tag{4a}$$
$$\frac{dE^*(t)}{dt} = q(t)b(t)\exp(-E^*(t)) + \beta(t)\exp(S^*(t)) - \alpha(t) - \chi(t) - \mu(t), \tag{4b}$$
$$\frac{dI^*(t)}{dt} = \alpha(t)\exp((E^* - I^*)(t)) - \delta(t)\exp(K^*(t)) - \gamma(t) - \varepsilon(t) - \mu(t), \tag{4c}$$
$$\frac{dK^*(t)}{dt} = \delta(t)\exp(I^*(t)) + \gamma(t)\exp((I^* - K^*)(t)) + \chi(t)\exp((E^* - K^*)(t)) - \mu(t). \tag{4d}$$
$$\frac{dR^*(t)}{dt} = \phi(t)\exp((K^* + S^* - R^*)(t)) + \varepsilon(t)\exp((I^* - R^*)(t)) - \mu(t). \tag{4e}$$

Thus, our aim is to study the positive periodic solutions of Equation (2) equivalently replaced by the analysis of positive periodic solution of the new system (4).

Theorem 1. *Consider the sets of functions $\{S, E, I, K, R\}$ and $\{S^*, E^*, I^*, K^*, R^*\}$ are related by Equation (3). Then, the functions S, E, I, K and R are a solution of the system in Equation (2) if and only if the functions S^*, E^*, I^*, K^* and R^* are a solution of the system in Equation (4). In particular, we have that the following two assertions are valid: (a) If S^*, E^*, I^*, K^* and R^* satisfying the system in Equation (4) are ω-periodic functions, then the functions S, E, I, K and R satisfying the system in Equation (2) are ω-periodic; and (b) The existence of a solution for the system in Equation (4) imply the existence of a positive solution for the system in Equation (2).*

Proof. The proof fact that $\{S, E, I, K, R\}$ is a solution of the system in Equation (2) if and only if $\{S^*, E^*, I^*, K^*, R^*\}$ is straightforward by the change of variable (3), differentiation and algebraic rearrangements. Now, we get the proof of item (a) by using the change of variable (3), for illustration, we consider the case of function S and we have that $S(t + \omega) = \exp(S^*(t + \omega)) = \exp(S^*(t)) = S(t)$. The item (b) is a straightforward consequence of the definition of the functions S^*, E^*, I^*, K^* and R^* given in Equation (3). □

In order to define the operator equation, we consider the normed vector spaces X and Y and introduce the operators $L : \text{Dom}\, L \subset X \to Y$ and $N : X \to Y$ explicitly defined by the relations

$$L\left((x_1, x_2, x_3, x_4, x_5)^T\right) = \left(\frac{dx_1}{dt}, \frac{dx_2}{dt}, \frac{dx_3}{dt}, \frac{dx_4}{dt}, \frac{dx_5}{dt}\right)^T \tag{5}$$
$$N\left((x_1, x_2, x_3, x_4, x_5)^T\right) = (\mathcal{N}_1, \mathcal{N}_2, \mathcal{N}_3, \mathcal{N}_4, , \mathcal{N}_5)^T, \tag{6}$$

where

$$\mathcal{N}_1(t) = p(t)b(t)\exp(-x_1(t)) - \beta(t)\exp(x_2(t)) - \phi(t)\exp((x_4 - x_1)(t)) - \mu(t), \tag{7}$$
$$\mathcal{N}_2(t) = q(t)b(t)\exp(-x_2(t)) + \beta(t)\exp(x_1(t)) - \alpha(t) - \chi(t) - \mu(t), \tag{8}$$
$$\mathcal{N}_3(t) = \alpha(t)\exp((x_2 - x_3)(t)) - \delta(t)\exp(x_4(t)) - \gamma(t) - \varepsilon(t) - \mu(t), \tag{9}$$
$$\mathcal{N}_4(t) = \delta(t)\exp(x_3(t)) + \gamma(t)\exp((x_3 - x_4)(t)) + \chi(t)\exp((x_2 - x_4)(t)) - \mu(t), \tag{10}$$
$$\mathcal{N}_5(t) = \phi(t)\exp((x_1 + x_4 - x_5)(t)) + \varepsilon(t)\exp((x_3 - x_5)(t)) - \mu(t). \tag{11}$$

The operator notation implies that the system in Equation (4) can be rewritten as the following operator equation

$$L\left((S^*, E^*, I^*, K^*, R^*)^T\right) = N\left((S^*, E^*, I^*, K^*, R^*)^T\right), \quad (S^*, E^*, I^*, K^*, R^*) \in \text{Dom}\, L \subset X, \tag{12}$$

where the appropriate Banach spaces X and Y are defined by

$$X = Y = \left\{ \mathbf{x}^T \in C(\mathbb{R}, \mathbb{R}^5) \ : \ \mathbf{x}(t+\omega) = \mathbf{x}(t), \quad \left\| \mathbf{x} \right\| = \sum_{i=1}^{5} \max_{t\in[0,\omega]} |x_i(t)| < \infty \right\}. \tag{13}$$

Hereinafter we use the bold notation $\mathbf{x}^T := (x_1, x_2, x_3, x_4, x_5)^T$. We notice that the spaces in Equation (13) are the more convenient, since we are concerned with the analysis of ω-periodic solutions. However, if the interest is to analyze other properties we should be consider a suitable definition of X and Y.

1.4. Main Results

By convenience of presentation, we introduce the notation

$$\overline{f} = \frac{1}{\omega}\int_0^{\omega} f(t)dt, \quad f^{\perp} = \min_{x\in[0,\omega]} f(x), \quad \text{and} \quad f^{\top} = \max_{x\in[0,\omega]} f(x), \tag{14}$$

for any positive real valued bounded function f defined on $[0, \omega]$.

Let us consider the following assumption

$$\left.\begin{array}{l} \text{The initial condition } (S(0), E(0), I(0), K(0), R(0)) \in \mathbb{R}^5_+\text{; the coefficient functions } b, p, \\ q, \alpha, \beta, \gamma, \chi, \phi, \delta, \mu \text{ and } \varepsilon \text{ are positive, continuous, } \omega\text{-periodic on } [0,\omega]\text{; and there are the} \\ \text{strictly positive constants } \kappa_1 \text{ and } \kappa_2 \text{ such that} \\ (pb)^{\perp} - \phi^{\top}(\phi^{\perp})^{-1}\overline{b}\exp(2\omega\overline{\mu}) \geq \kappa_1 > 0 \\ 1 - \dfrac{\varepsilon^{\top}\alpha^{\top}\overline{b}}{\mu^{\top}(\gamma+\varepsilon+\mu)^{\perp}(\alpha+\chi+\mu)^{\perp}} \exp\left(\omega\left[(\gamma+\varepsilon+\mu)^{\top} + \phi^{\top}(\phi^{\perp})^{-1}\overline{b} + \mu^{\top}\right]\right) \geq \kappa_2 > 0 \end{array}\right\} \tag{15}$$

Then, the main result of the paper are given by the following three theorems.

Theorem 2. *Let X and Y the spaces defined on Equation (13); $Q : Y \to Y$ defined by $Q(\mathbf{x}^T) = \omega^{-1}\int_0^{\omega} \mathbf{x}(\tau)^T d\tau$; and the operators $L : X \to Y$ and $N : X \to Y$ defined on Equations (5) and (6), respectively. Moreover, assume that the hypothesis in Equation (15) is satisfied. Then, there are the positive constants $\rho_1, \rho_2, \rho_3, d_1, d_2, d_3, \delta_1, \delta_2$ and δ_3, such that the following two assertions are valid*

(a) *If $\lambda \in]0,1[$ and $\mathbf{x} \in \operatorname{Dom} L$ are such that $L(\mathbf{x}) = \lambda N(\mathbf{x})$, the following inequalities*

$$x_i(t) < \ln(\rho_i/\omega) + d_i, \quad i = 1, \ldots, 5 \tag{16}$$
$$\ln(\delta_i) < x_i(t), \quad i = 1, \ldots, 5, \tag{17}$$

holds for all $t \in [0, \omega]$.

(b) *If $\mathbf{x} \in \operatorname{Ker} L$ are such that $QN\mathbf{x} = 0$, the following inequalities*

$$x_i(t) < \ln(\rho_i/\omega), \quad i = 1, \ldots, 5, \tag{18}$$
$$\ln(\delta_i) < x_i(t), \quad i = 1, \ldots, 5, \tag{19}$$

holds for all $t \in [0, \omega]$.

Theorem 3. *If the hypothesis in Equation (15) is satisfied, there exists at least one ω-periodic solution of Equation (4).*

Theorem 4. *Consider that the hypothesis in Equation (15) is satisfied. Then, the system in Equation (2) has at least one positive ω-periodic solution.*

1.5. Related Works

There are several works where the study of positive periodic solutions is developed, for instance in [13,19–33]. In particular, recently in [13] was proved the existence of at one positive periodic solution of the following system modeling the dynamics of a computer virus

$$\begin{aligned}
\frac{dS(t)}{dt} &= b(t) - \mu_1(t)S(t) - \beta_1(t)S(t)L(t) - \beta_2(t)S(t)A(t) + \gamma_1(t)L(t) + \gamma_2(t)A(t), \\
\frac{dL(t)}{dt} &= \beta_1(t)S(t)L(t) + \beta_2(t)S(t)A(t) + \alpha_2(t)A(t) - [\mu_2(t) + \alpha_1(t) + \gamma_1(t)]L(t), \\
\frac{dA(t)}{dt} &= \alpha_1(t)L(t) - [\mu_3(t) + \alpha_1(t) + \gamma_2(t)]A(t),
\end{aligned}$$

by assuming that $S(0), L(0)$ and $A(0)$ are strictly positive and the functions $b, \mu_1, \mu_2, \mu_3, \beta_1, \beta_2, \gamma_1, \gamma_2, \alpha_1$ and α_2 are positive, continuous, ω-periodic on $[0, \omega]$ and

$$\left(\frac{\alpha_1}{\alpha_1 + \mu_2}\right)^{\top} (\alpha_2 + \gamma_2)^{\top} < (\mu_3 + \alpha_2 + \gamma_2)^{\perp}$$

We observe that S, L and A denotes the susceptible computers, the latent computers and the infectious computers, respectively.

1.6. Outline of the Paper

The paper is organized as follows. In Section 2, we introduce some terminology related to the coincidence degree theory and some useful results. In Sections 3–5 we develop the proof of Theorems 2–4, respectively. Finally, in Section 6, we present an examples of a system with coefficients satisfying Equation (15).

2. Preliminaries

In this paper, we utilize the standard notation and terminology of topological degree theory. However, for self-contained presentation, we recall some notation, concepts and results related to the statement of of Mawhin's theorem, [34]. Moreover, we prove some properties for the operators L and N defining on the operator Equation (12).

2.1. The Mawhin's Continuation Theorem

Definition 1. *Let X and Y be normed vector spaces and $L : \text{Dom } L \subset X \to Y$ a linear operator. Then, L is called a Fredholm operator of index zero, if the following assertions*

$$\dim(\text{Ker } L) = \text{codim}(\text{Im } L) < \infty \qquad \textit{and} \qquad \text{Im } L \textit{ is closed in } Y, \tag{20}$$

are valid.

Proposition 1. *Let X and Y be normed vector spaces and $L : \text{Dom } L \subset X \to Y$ a linear operator. If L is a Fredholm mapping of index zero, then*

(i) There are two continuous projectors $P : X \to X$ and $Q : Y \to Y$ such that $\text{Im } P = \text{Ker } L$ and $\text{Im } L = \text{Ker } Q = \text{Im } (I - Q)$.
(ii) $L_P := L|_{\text{Dom } L \cap \text{Ker } P} : (I - P)X \to \text{Im } L$ is invertible and its inverse is denoted by K_P.
(iii) There is an isomorphism $J : \text{Im } Q \to \text{Ker } L$.

Definition 2. *Let X and Y be normed vector spaces and $L : \text{Dom } L \subset X \to Y$ a Fredholm mapping of index zero. Let $P : X \to X$ and $Q : Y \to Y$ be two continuous projectors such that $\text{Im } P = \text{Ker } L$ and $\text{Im } L = \text{Ker } Q = \text{Im } (I - Q)$. Let us consider $N : X \to Y$ a continuous operator and $\Omega \subset X$ an open* *bounded set. Then, N is called L−compact on $\overline{\Omega}$ if $QN(\overline{\Omega})$ is a bounded set and the operator $K_P(I-Q)N$ is compact on $\overline{\Omega}$.*

Definition 3. *Let $\Omega \subset \mathbb{R}^n$ be an open bounded set, $f \in C^1(\Omega,\mathbb{R}^n) \cap C(\overline{\Omega},\mathbb{R}^n)$ and $y \in \mathbb{R}^n \backslash f(\partial\Omega \cup N_f)$, i.e., y is a regular value of f. Here, $N_f = \{x \in \Omega : J_f(x) = 0\}$ the critical set of f and J_f the Jacobian of f at x. Then, the degree $\deg\{f,\Omega,y\}$ is defined by $\deg\{f,\Omega,y\} = \sum_{x\in f^{-1}(y)} \operatorname{sgn} J_f(x)$, with the agreement that $\sum_{\emptyset} = 0$.*

Theorem 5. *Assume that $(X,\|.\|_X)$ and $(Y,\|.\|_Y)$ are two Banach spaces and Ω is an open bounded set. Consider that $L : \text{Dom } L \subset X \to Y$ be a Fredholm mapping of index zero and $N : X \to Y$ be L-compact on $\overline{\Omega}$. If the following hypotheses*

(C_1) $Lx \neq \lambda Nx$ for each $(\lambda, x) \in]0,1[\times(\partial\Omega \cap \text{Dom } L)$.
(C_2) $QNx \neq 0$ for each $x \in \partial\Omega \cap \text{Ker } L$.
(C_3) $deg(JQN, \Omega \cap \text{Ker } L, 0) \neq 0$.

are valid. Then the operator equation $Lx = Nx$ has at least one solution in $\text{Dom } L \cap \overline{\Omega}$.

2.2. L Is a Fredholm Operator of Index Zero

Lemma 1. *The operator $L : \text{Dom } L \subset X \to Y$ defined on Equation (5), with X and Y the Banach spaces given on Equation (13), is a Fredholm operator of index zero. Moreover the sets Ker L and Im L are characterized by $\text{Ker } L \cong \mathbb{R}^5$ and $\text{Im } L = \left\{\mathbf{y} \in Y : \int_0^\omega \mathbf{y}(\tau)^T d\tau = 0\right\}$, respectively.*

Proof. In order to prove the Lemma we apply the Definition 1 or more precisely we prove that L satisfy Equation (20).

The left condition in Equation (20) is proved as follows. Let $(s_0, l_0, i_0, k_0, r_0) \in \mathbb{R}^5$ such that $\mathbf{x}(t_0) = (s_0, l_0, i_0, k_0, r_0)$, we observe that $\mathbf{x}^T \in \text{Ker } L$ is equivalent to $\mathbf{x}(t) = (s_0, l_0, i_0, k_0, r_0)$ for all $t \geq t_0$. Then, we have that $\text{Ker } L \cong \mathbb{R}^5$. Now, if we select arbitrarily $\mathbf{y}^T \in \text{Im } L$, we have that there is $\mathbf{x} \in \text{Dom } L$ such that $L\mathbf{x}^T = \mathbf{y}^T$. Then, from Equation (5) and ω-periodic behavior of $\mathbf{x}$, we deduce that $\int_t^{t+\omega} \mathbf{y}(\tau)^T d\tau = 0$ for each $t \geq t_0$ or equivalently $\text{Im } L = \left\{\mathbf{y} \in Y : \int_0^\omega \mathbf{y}(\tau)^T d\tau = 0\right\}$. Now, by linear algebra results, we recall the existence of isomorphisms $X \cong \text{Im } L \oplus (X/\text{Im } L)$, $X \cong \text{Ker } L \oplus (X/\text{Ker } L)$, and $\text{Im } L \cong X/\text{Ker } L$. Thus, we have that $\text{Ker } L \cong X/\text{Im } L$ and we get that $\dim(\text{Ker } L) = \text{codim}(\text{Im } L) = 4$.

To prove the left condition in Equation (20) we introduce the linear continuous mapping $F : \text{Im } L \subset Y \to \mathbb{R}^5$ defined by $F(\mathbf{x}^T) = \int_0^\omega \mathbf{x}^T(\tau)d\tau$ and observe that $F^{-1}(0) = \text{Im } L$. Thus, clearly $\text{Im } L$ is a closed set of the space Y. □

2.3. Construction of the Projectors P, Q and the Operator K_P

We remark that the existence of three abstract projectors P, Q and K_P associated to L, is guaranteed by Proposition 1. However, by convenience of some calculus in the following sections we introduce explicitly the definitions of P and Q given by

$$P : X \to X, \quad Q : Y \to Y, \quad P\left(\mathbf{x}^T\right) = Q\left(\mathbf{x}^T\right) = \frac{1}{\omega}\int_0^\omega \mathbf{x}(\tau)^T d\tau \tag{21}$$

and notice that satisfy the relations in Proposition 1. More precisely, we have that

(a) <u>$\text{Ker } L = \text{Im } P$</u>. We prove that $\text{Ker } L \subset \text{Im } P$ as follows: from the isomorphism $\text{Ker } L \cong \mathbb{R}^5$ given on Lemma 1, we observe that $\mathbf{x}^T \in \text{Ker } L$ is equivalent to the fact that $\mathbf{x}(t)$ is constant for all $t \geq t_0$, which at the same time implies that $\mathbf{x} \in \text{Im } P$, since for $\mathbf{x}(t)$ constant we have that $P\left(\mathbf{x}^T\right) = \mathbf{x}^T$ Conversely, the proof of the inclusion $\text{Im } P \subset \text{Ker } L$ is deduced by the following facts: for $\mathbf{y}^T \in \text{Im } P$ there is $\mathbf{z} \in X$ such that $P(\mathbf{z}^T) = \mathbf{y}^T$ and from Equation (21) we obtain that $\omega^{-1} \int_0^\omega \mathbf{z}(\tau)^T d\tau = \mathbf{y}^T$ which implies by differentiation the fact that $L(\mathbf{y}^T) = 0$ or $\mathbf{y} \in \text{Ker } L$.

(b) Ker $Q = \text{Im } L$. From the definition of Q given in Equation (21) we have that $\mathbf{y}^T \in \text{Ker } Q$ is equivalent to $\int_0^\omega \mathbf{y}(\tau)^T d\tau = 0$ and from the characterization of Im L given on Lemma 1 is at the same time equivalent to $\mathbf{y}^T \in \text{Im } L$.

(c) Im $(I - Q) = \text{Im } L$. Let $\mathbf{y}^T \in \text{Im } (I - Q)$, then there is $\mathbf{z} \in X$ such that $(I - Q)\left(\mathbf{z}^T\right) = \mathbf{y}^T$, which implies that

$$\int_0^\omega \mathbf{y}(\tau)^T d\tau = \int_0^\omega \left(\mathbf{z}(\tau)^T - \frac{1}{\omega} \int_0^\omega \mathbf{z}(m)^T dm \right) d\tau = (0,0,0,0)$$

and, from the characterization of Im L given on Lemma 1, we get that $\mathbf{y}(\tau)^T \in \text{Im } L$. Thus, we obtain that Im $(I - Q) \subset \text{Im } L$. By analogous arguments, we can prove the inclusion Im $L \subset \text{Im } (I - Q)$.

(d) Operators K_P and L_P. The notation L_P is is introduced for the restriction of L to Dom $L \cap \text{Ker } P$, i.e., L_P is the operator defined from Dom $L \cap \text{Ker } P$ to Im L and $L_P = L$ on Dom $L \cap \text{Ker } P$. The symbol K_P is used to denote the inverse of L_P, and is precisely defined as the operator such that

$$K_P\left(\mathbf{x}^T\right)(t) = \int_0^t \mathbf{x}(\tau)^T d\tau - \frac{1}{\omega} \int_0^\omega \int_0^\eta \mathbf{x}(m)^T dm d\eta. \tag{22}$$

We notice that, we can prove that the operator K_P is the inverse of the operator L_P by application of the following identity

$$\int_0^t \frac{d}{ds}\mathbf{x}(s) ds - \frac{1}{\omega} \int_0^\omega \int_0^t \frac{d}{dm}\mathbf{x}(m) dm dt = \mathbf{x}(t),$$

which is valid only for all $\mathbf{x}^T \in \text{Dom } L \cap \text{Ker } P$.

Thus, the projectors P and Q defined on Equation (21) satisfy the Proposition 1, since we can follow (i) and (ii) are satisfied from (a)–(c) and (d), respectively.

2.4. *N Defined on Equation (6) Is a Continuous Operator*

Lemma 2. *The operator $N : X \to Y$ defined on Equation (5), with X and Y the Banach spaces given on Equation (13), is a continuous operator.*

Proof. Let us choose arbitrarily the sequence $\{\mathbf{x}_n\} \subset X$ which converges to $\overline{\mathbf{x}}$ in the norm induced topology of X. By the definition of N given on Equation (6) and applying componentwise the inequality

$$|\exp(z_2) - \exp(z_1)| = \left| \int_{z_2}^{z_1} \exp(s) ds \right| \le \max\left\{ \exp(z_1), \exp(z_2) \right\} |z_2 - z_1|, \quad \forall z_1, z_2 \in \mathbb{R},$$

we get the existence of $C > 0$ depending only on $b, \mu_1, \beta_1, \beta_2, \gamma_1, \gamma_2, \alpha_1$ and α_2 such that $\|N(\mathbf{x}_n) - N(\overline{\mathbf{x}})\| \le C\|\mathbf{x}_n - \overline{\mathbf{x}}\|$. Thus, the sequence $\{N(\mathbf{x}_n)\} \subset X$ converges to $N(\overline{\mathbf{x}})$ in the topology of X induced by the norm. Hence, we can deduce that N is a continuous operator. □

2.5. *N Defined on Equation (6) Is L-Compact on any Ball of X Centered at* $(0,0,0,0,0)$.

Lemma 3. *Assuming that $h \in \mathbb{R}^+$ is an arbitrary and fix number defining the radius h of an open ball of X centered at $(0,0,0,0,0)$, denoted by $\Omega \subset X$, i.e.,*

$$\Omega = \left\{ (x_1, x_2, x_3, x_4, x_5) \in X \quad : \quad \|(x_1, x_2, x_3, x_4, x_5)\| < h \right\}. \tag{23}$$

Moreover, consider L and N defined on Eqautions (5) and (6), respectively. If the assumption in Equation (15) is satisfied, the operator N is L-compact on $\overline{\Omega}$.

Proof. The proof is focused in the verification of the fact that L satisfy the two requirements of Definition 2: $QN(\overline{\Omega})$ is a bounded set and $K_P(I-Q)N$ is a compact operator on $\overline{\Omega}$, since Ω is an open bounded set by the its definition given on Equation (23) and L is a Fredholm operator of index zero by application of Lemma 1.

To prove that $QN(\overline{\Omega})$ is bounded we proceed as follows. We observe that

$$QN(\mathbf{x}^T) = \frac{1}{\omega}\int_0^{\omega} N(\tau)^T d\tau. \tag{24}$$

Then, for $\mathbf{x} \in \overline{\Omega}$ we have that $\|QN(\mathbf{x}^T)\| \leq \frac{1}{\omega}\int_0^{\omega} \|N\| d\tau = \|N\|$, which implies that $QN(\overline{\Omega})$ is bounded.

In order to prove that $K_P(I-Q)N$ is a compact operator on $\overline{\Omega}$, we observe that from Equations (6), (21) and (22) we get

$$(K_P(I-Q)N)(\mathbf{x}^T)(t) = \int_0^t N(\tau)^T d\tau + \left(\frac{1}{2} - \frac{t}{\omega}\right)\int_0^{\omega} N(\tau)^T d\tau - \frac{1}{\omega}\int_0^{\omega}\int_0^{\eta} N(m)^T dm d\eta.$$

Then, we deduce that $\|K_P(I-Q)N\| \leq 2\omega\|N\|$, as a result we have that $(K_P(I-Q)N)(\overline{\Omega})$ is a bounded, since the operator N is bounded on $\overline{\Omega}$. Moreover, we can prove the bound

$$|(K_P(I-Q)N)(\mathbf{x}^T)(t) - (K_P(I-Q)N)(\mathbf{x}^T)(s)| \leq 2\|N\|\,|t-s|, \quad \forall t,s \in [t_0,\infty[,$$

i.e., $K_P(I-Q)N$ is an equicontinuous operator. Hence, by Arzela Ascoli's theorem we get that $K_P(I-Q)N$ is a compact operator on $\overline{\Omega}$. □

2.6. A Useful Auxiliary Result

Proposition 2. *[13] Let* $\psi : [0,\omega] \subset \mathbb{R}^+ \to \mathbb{R}$ *be an absolutely continuous function satisfying the differential inequality*

$$\frac{d}{dt}\psi(t) + m(t)\psi(t) \geq 0, \quad \forall t \in [0,\omega], \tag{25}$$

with $m \in L^1([0,\omega])$ *such that* $0 < m_1 \leq m(t) \leq m_2$ *for all* $t \in [0,\omega]$ *and for some positive constants* m_1 *and* m_2. *Then, if* $\psi(0) > 0$ *we have that* $\psi(t) \geq \psi(0)\exp(-m_2\omega) > 0$ *for all* $t \in [0,\omega]$.

3. Proof of Theorem 2

3.1. Four Useful Lemmata

We introduce four Lemmmata related with some estimates for the operator equation $L\mathbf{x} = \lambda N\mathbf{x}$, which is equivalent to the following system

$$\frac{dx_1}{dt} = \lambda\left[pb\exp(-x_1) - \beta\exp(x_2) - \phi\exp(x_4 - x_1) - \mu\right], \tag{26a}$$
$$\frac{dx_2}{dt} = \lambda\left[qb\exp(-x_2) + \beta\exp(x_1) - \alpha - \chi - \mu\right], \tag{26b}$$
$$\frac{dx_3}{dt} = \lambda\left[\alpha\exp(x_2 - x_3) - \delta\exp(x_4) - \gamma - \varepsilon - \mu\right], \tag{26c}$$
$$\frac{dx_4}{dt} = \lambda\left[\delta\exp(x_3) + \gamma\exp(x_3 - x_4) + \chi\exp(x_2 - x_4) - \mu\right], \tag{26d}$$
$$\frac{dx_5}{dt} = \lambda\left[\phi\exp(x_1 + x_4 - x_5) + \varepsilon\exp(x_3 - x_4) - \mu\right], \tag{26e}$$

and also can be rewritten as the system

$$\frac{d}{dt}\exp(x_1) + \lambda\mu\exp(x_1) = \lambda\left[pb - \beta\exp(x_1 + x_2) - \phi\exp(x_4)\right], \tag{27a}$$
$$\frac{d}{dt}\exp(x_2) + \lambda(\alpha + \chi + \mu)\exp(x_2) = \lambda\left[qb + \beta\exp(x_1 + x_2)\right], \tag{27b}$$
$$\frac{d}{dt}\exp(x_3) + \lambda(\gamma + \varepsilon + \mu)\exp(x_3) = \lambda\left[\alpha\exp(x_2) - \delta\exp(x_3 + x_4)\right], \tag{27c}$$
$$\frac{d}{dt}\exp(x_4) + \lambda\mu\exp(x_4) = \lambda\left[\delta\exp(x_3 + x_4) + \gamma\exp(x_3) + \chi\exp(x_2)\right], \tag{27d}$$
$$\frac{d}{dt}\exp(x_5) + \lambda\mu\exp(x_5) = \lambda\left[\phi\exp(x_1 + x_4) + \varepsilon\exp(x_3 - x_4 + x_5)\right]. \tag{27e}$$

We notice that to deduce Equation (27) we multiply the i-th equation of the system in Equation (26) by $\exp(x_i)$. Thus, the proof of estimates for $L\mathbf{x} = \lambda N\mathbf{x}$ is focused in to get the estimates of the solutions of Equation (26) (or equivalently of Equation (27)).

Lemma 4. *Assume that $(S(0), E(0), I(0), K(0), R(0)) \in \mathbb{R}^5_+$; the coefficient functions $b, p, q, \alpha, \beta, \gamma, \chi, \phi, \delta, \mu$ and ε are positive, continuous and ω-periodic on $[0, \omega]$; and the operators $L : \mathrm{Dom}\, L \subset X \to Y$ and defined on Equations (5) and (6), with X and Y the Banach spaces given on Equation (13). Then, the solution of the operator equation $L\mathbf{x} = \lambda N\mathbf{x}$ with $\lambda \in]0, 1[$ satisfy the following inequalities*

$$\exp(x_2(t)) \geq \exp\left(E(0) - (\alpha + \chi + \mu)^{\top}\omega\right), \tag{28}$$
$$\exp(x_4(t)) \geq \exp\left(K(0) - \mu^{\top}\omega\right), \tag{29}$$
$$\exp(x_5(t)) \geq \exp\left(R(0) - \mu^{\top}\omega\right), \tag{30}$$
$$(pb)^{\perp} \leq \left[\mu^{\top} + \beta^{\top}\max_{t\in[0,\omega]}\exp(x_2(t))\right]\exp(x_1(t)) + \phi^{\top}\max_{t\in[0,\omega]}\exp(x_4(t)), \tag{31}$$
$$\alpha^{\perp}\exp\left(E(0) - (\alpha + \chi + \mu)^{\top}\omega\right) \leq \left[(\gamma + \varepsilon + \mu)^{\top} + \delta^{\top}\max_{t\in[0,\omega]}\exp(x_4(t))\right]\exp(x_3(t)), \tag{32}$$

for any $t \in [0, \omega]$.

Proof. By the continuity of the coefficient functions and the fact that $\lambda \in]0, 1[$, we have that $\lambda(\alpha + \chi + \mu)(t) \in [\lambda(\alpha + \chi + \mu)^{\perp}, (\alpha + \chi + \mu)^{\top}] \subset \mathbb{R}^+$ and $\lambda\mu(t) \in [\lambda\mu^{\perp}, \mu^{\top}] \subset \mathbb{R}^+$, for any $t \in [0, \omega]$. Then, we can prove Equations (28)–(30), by straightforward application of Proposition 2 to Equations (27)b,d,e, respectively, since we have that

$$\exp(x_2(t)) \geq \exp(x_2(0)) \exp(-(\alpha+\chi+\mu)^\top \omega) = \exp\left(E(0) - (\alpha+\chi+\mu)^\top \omega\right),$$
$$\exp(x_4(t)) \geq \exp(x_4(0)) \exp(-\mu^\top \omega) = \exp\left(K(0) - \mu^\top \omega\right),$$
$$\exp(x_5(t)) \geq \exp(x_5(0)) \exp(-\mu^\top \omega) = \exp\left(R(0) - \mu^\top \omega\right),$$

for any $t \in [0,\omega]$. Now, to prove Equations (31) and (32), for $i = 1,3$, we introduce the notation $\tau_i \in [0,\omega]$ for the points where x_i has a minimum. Then, using the notation in Equation (14), from Equations (27)a,c, and (28) we get

$$\begin{aligned}
(pb)^\perp &\leq (pb)(\tau_1) \\
&= \mu(\tau_1)\exp(x_1(\tau_1)) + \beta(\tau_1)\exp((x_1+x_2)(\tau_1)) + \phi(\tau_1)\exp(x_4(\tau_1)) \\
&= \left[\mu(\tau_1) + \beta(\tau_1)\exp(x_2(\tau_1))\right]\exp(x_1(\tau_1)) + \phi(\tau_1)\exp(x_4(\tau_1)) \\
&\leq \left[\mu^\top + \beta^\top \max_{t\in[0,\omega]} \exp(x_2(t))\right]\exp(x_1(t)) + \phi^\top \max_{t\in[0,\omega]} \exp(x_4(t)), \\
\alpha^\perp \exp\left(E(0) - (\alpha+\chi+\mu)^\top\omega\right) &\leq \alpha(\tau_3)\exp(x_2(\tau_3)) \\
&= (\gamma+\varepsilon+\mu)(\tau_3)\exp(x_3(\tau_3)) + \delta(\tau_3)\exp((x_3+x_4)(\tau_3)) \\
&\leq \left[(\gamma+\varepsilon+\mu)^\top + \delta^\top \max_{t\in[0,\omega]}\exp(x_4(t))\right]\exp(x_3(t)),
\end{aligned}$$

for any $t \in [0,\omega]$. □

Lemma 5. *Assume that hypotheses of Lemma 4. Then, the solution of the operator equation* $L\mathbf{x} = \lambda N\mathbf{x}$ *with* $\lambda \in]0,1[$ *satisfy the integral inequalities*

$$\int_0^\omega \exp(x_1(t))dt \leq \frac{\omega\bar{b}}{\mu^\perp}, \tag{33}$$
$$\int_0^\omega \exp(x_2(t))dt \leq \frac{\omega\bar{b}}{(\alpha+\chi+\mu)^\perp}, \tag{34}$$
$$\int_0^\omega \exp(x_3(t))dt \leq \frac{\omega\alpha^\top\bar{b}}{(\alpha+\chi+\mu)^\perp(\gamma+\varepsilon+\mu)^\perp}, \tag{35}$$
$$\int_0^\omega \exp(x_4(t))dt \leq \frac{\omega\bar{b}}{\phi^\perp}, \tag{36}$$
$$\int_0^\omega \exp(x_5(t))dt \leq \frac{\omega\bar{b}\phi^\top}{\mu^\perp\phi^\perp}\max_{t\in[0,\omega]}\exp(x_1(t)) + \frac{\varepsilon^\top \max_{t\in[0,\omega]}\exp(x_3(t))}{\mu^\perp \exp(K(0)-\mu^\top\omega)}\int_0^\omega \exp(x_5(t))dt. \tag{37}$$

Proof. We integrate the equations of the system in Equation (27) on $[0, \omega]$ and using the ω-periodicity of $\mathbf{x}$ we deduce the following identities

$$\int_0^\omega p(t)b(t)dt = \int_0^\omega [\beta(t)\exp((x_1 + x_2)(t)) + \phi(t)\exp(x_4(t)) + \mu(t)\exp(x_1(t))]\,dt, \quad (38a)$$

$$\int_0^\omega q(t)b(t)dt = \int_0^\omega [-\beta(t)\exp((x_1 + x_2)(t)) + (\alpha + \chi + \mu)(t)\exp(x_2(t))]\,dt, \quad (38b)$$

$$\int_0^\omega \alpha(t)\exp(x_2(t))dt = \int_0^\omega [\delta(t)\exp((x_3 + x_4)(t)) + (\gamma + \varepsilon + \mu)(t)\exp(x_3(t))]\,dt, \quad (38c)$$

$$\int_0^\omega \mu(t)\exp(x_4(t))dt = \int_0^\omega [\delta(t)\exp((x_3 + x_4)(t)) + \gamma(t)\exp(x_3(t)) + \chi(t)\exp(x_2(t))]\,dt, \quad (38d)$$

$$\int_0^\omega \mu(t)\exp(x_5(t))dt = \int_0^\omega [\phi(t)\exp((x_1 + x_4)(t)) + \varepsilon(t)\exp((x_3 - x_4 + x_5)(t))]\,dt. \quad (38e)$$

Then, adding Equation (38)a,b, using the ω-periodicity of x_1 and x_2, and the fact that $p(t) + q(t) = 1$, we deduce that

$$\int_0^\omega b(t)dt = \int_0^\omega [\mu(t)\exp(x_1(t)) + \{\alpha(t) + \chi(t) + \mu(t)\}\exp(x_2(t)) + \phi(t)\exp(x_4(t))]\,dt,$$

which implies Equations (33), (34) and (36), since, by the positivity of α, χ, μ and ϕ and the notation in Equation (14), we get the inequalities

$$\begin{aligned}
\int_0^\omega \exp(x_1(t))dt &\leq \frac{1}{\min\limits_{t\in[0,\omega]} \mu(t)} \int_0^\omega \mu(t)\exp(x_1(t))dt \\
&\leq \frac{1}{\mu^\perp} \int_0^\omega [\mu(t)\exp(x_1(t)) + \{\alpha(t) + \chi(t) + \mu(t)\}\exp(x_2(t)) + \phi(t)\exp(x_4(t))]\,dt \\
&= \frac{1}{\mu^\perp} \int_0^\omega b(t)dt = \frac{\omega\bar{b}}{\mu^\perp}, \\
\int_0^\omega \exp(x_2(t))dt &\leq \frac{1}{\min\limits_{t\in[0,\omega]} (\alpha + \chi + \mu)(t)(t)} \int_0^\omega (\alpha + \chi + \mu)(t)\exp(x_2(t))dt \\
&\leq \frac{1}{(\alpha + \chi + \mu)^\perp} \int_0^\omega [\mu(t)\exp(x_1(t)) + \{\alpha(t) + \chi(t) + \mu(t)\}\exp(x_2(t)) + \phi(t)\exp(x_4(t))]\,dt \\
&= \frac{1}{(\alpha + \chi + \mu)^\perp} \int_0^\omega b(t)dt = \frac{\omega\bar{b}}{(\alpha + \chi + \mu)^\perp}, \\
\int_0^\omega \exp(x_4(t))dt &\leq \frac{1}{\min\limits_{t\in[0,\omega]} \phi(t)} \int_0^\omega \phi(t)\exp(x_4(t))dt \\
&\leq \frac{1}{\phi^\perp} \int_0^\omega [\mu(t)\exp(x_1(t)) + \{\alpha(t) + \chi(t) + \mu(t)\}\exp(x_2(t)) + \phi(t)\exp(x_4(t))]\,dt \\
&= \frac{1}{\phi^\perp} \int_0^\omega b(t)dt = \frac{\omega\bar{b}}{\phi^\perp}.
\end{aligned}$$

The inequality in Equation (35) is a consequence of Equations (38)c and (34), since

$$\begin{aligned}
\int_0^\omega \exp(x_3(t))dt &\leq \frac{1}{\min\limits_{t\in[0,\omega]}(\gamma+\varepsilon+\mu)(t)}\int_0^\omega (\gamma+\varepsilon+\mu)(t)\exp(x_3(t))dt \\
&\leq \frac{1}{(\gamma+\varepsilon+\mu)^\perp}\int_0^\omega [\delta(t)\exp((x_3+x_4)(t))+(\gamma+\varepsilon+\mu)(t)\exp(x_3(t))]\,dt \\
&= \frac{1}{(\gamma+\varepsilon+\mu)^\perp}\int_0^\omega \alpha(t)\exp(x_2(t))dt \\
&\leq \frac{\omega\alpha^\top\overline{b}}{(\gamma+\varepsilon+\mu)^\perp(\alpha+\chi+\mu)^\perp}.
\end{aligned}$$

Now, from Equations (38)e, (36) and (32), we deduce the following estimate

$$\begin{aligned}
\mu^\perp\int_0^\omega \exp(x_5(t))dt &\leq \int_0^\omega \mu(t)\exp(x_5(t))dt \\
&= \int_0^\omega \Big[\phi(t)\exp((x_1+x_4)(t))+\varepsilon(t)\exp((x_3-x_4+x_5)(t))\Big]dt \\
&\leq \frac{\omega\overline{b}\phi^\top}{\phi^\perp}\max_{t\in[0,\omega]}\exp(x_1(t))+\varepsilon^\top\frac{\max_{t\in[0,\omega]}\exp(x_3(t))}{\min_{t\in[0,\omega]}\exp(x_4(t))}\int_0^\omega \exp(x_5(t))dt \\
&\leq \frac{\omega\overline{b}\phi^\top}{\phi^\perp}\max_{t\in[0,\omega]}\exp(x_1(t))+\varepsilon^\top\frac{\max_{t\in[0,\omega]}\exp(x_3(t))}{\exp(K(0)-\mu^\top\omega)}\int_0^\omega \exp(x_5(t))dt,
\end{aligned}$$

which implies Equation (37). □

Lemma 6. *Assume that hypotheses of Lemma 4. Then, the solution of the operator equation* $L\mathbf{x}=\lambda N\mathbf{x}$ *with* $\lambda\in]0,1[$ *satisfy the integral inequalities*

$$\int_0^\omega \left|\frac{dx_1}{dt}(t)\right|dt \leq 2\omega\overline{b}\max_{t\in[0,\omega]}\exp(-x_1(t)), \tag{39}$$

$$\int_0^\omega \left|\frac{dx_2}{dt}(t)\right|dt < 2\omega(\alpha+\chi+\mu)^\top, \tag{40}$$

$$\int_0^\omega \left|\frac{dx_3}{dt}(t)\right|dt < 2\omega\Big((\gamma+\varepsilon+\mu)^\top+\delta^\top\overline{b}(\phi^\top)^{-1}\Big), \tag{41}$$

$$\int_0^\omega \left|\frac{dx_4}{dt}(t)\right|dt < 2\omega\overline{\mu}, \tag{42}$$

$$\int_0^\omega \left|\frac{dx_5}{dt}(t)\right|dt < 2\omega\overline{\mu}\max_{t\in[0,\omega]}\exp(x_4(t)-x_5(t)). \tag{43}$$

Proof. We integrate the system in Equation (26) on $[0,\omega]$ and by using the ω-periodicity behavior of $\mathbf{x}$, we have that

$$\int_0^\omega p(t)b(t)\exp(-x_1(t))dt = \int_0^\omega [\beta(t)\exp(x_2(t))-\phi(t)\exp(x_4(t)-x_1(t))-\mu(t)]\,dt, \tag{44a}$$

$$\int_0^\omega [q(t)b(t)\exp(-x_2(t))+\beta(t)\exp(x_1(t))]\,dt = \int_0^\omega [\alpha(t)+\chi(t)+\mu(t)]\,dt, \tag{44b}$$

$$\int_0^\omega [\alpha(t)\exp(x_2(t)-x_3(t))-\delta(t)\exp(x_4(t))]\,dt = \int_0^\omega [\gamma(t)+\varepsilon(t)+\mu(t)]\,dt, \tag{44c}$$

$$\int_0^\omega [\delta(t)\exp(x_3(t))+\gamma(t)\exp(x_3(t)-x_4(t))+\chi(t)\exp(x_2(t)-x_4(t))]\,dt = \int_0^\omega \mu(t)dt, \tag{44d}$$

$$\int_0^\omega \phi(t)\exp(x_1(t)+x_4(t)-x_5(t))+\varepsilon(t)\exp(x_3(t)-x_4(t))-\mu(t)\exp(x_4(t)-x_5(t))dt = 0. \tag{44e}$$

Then, taking the modulus of the each equations defining the system in Equation (26); integrating each resulting equations on $[0,\omega]$; using the information that $\lambda \in]0,1[$; employing the relations of Equation (44); and applying the inequalities on Lemmas 4 and 5, we obtain the following estimates

$$
\begin{aligned}
\int_0^\omega \left|\frac{dx_1}{dt}(t)\right| dt &< 2\int_0^\omega p(t)b(t)\exp(-x_1(t))dt \le 2\omega\bar{b}\max_{t\in[0,\omega]}\exp(-x_1(t)),\\
\int_0^\omega \left|\frac{dx_2}{dt}(t)\right| dt &< 2\int_0^\omega (\alpha+\chi+\mu)(t)dt \le 2\omega(\alpha+\chi+\mu)^\top,\\
\int_0^\omega \left|\frac{dx_3}{dt}(t)\right| dt &< 2\int_0^\omega [(\gamma+\varepsilon+\mu)(t)+\delta(t)\exp(x_4(t))]\,dt\\
&\le 2\omega\Big((\gamma+\varepsilon+\mu)^\top+\delta^\top\bar{b}(\phi^\top)^{-1}\Big),\\
\int_0^\omega \left|\frac{dx_4}{dt}(t)\right| dt &< 2\int_0^\omega \mu(t)dt = 2\omega\bar{\mu},\\
\int_0^\omega \left|\frac{dx_5}{dt}(t)\right| dt &< 2\int_0^\omega \mu(t)\exp(x_4(t)-x_5(t))dt\\
&\le 2\omega\bar{\mu}\max_{t\in[0,\omega]}\exp(x_4(t)-x_5(t)),
\end{aligned}
$$

which conclude the proof of lemma. □

Lemma 7. *Assume that hypotheses of Lemma 4. Moreover consider that the hypotheses (15) and* $\mathbf{x}$ *is the solution of the operator equation* $\mathbf{L}\mathbf{x} = \lambda N\mathbf{x}$ *with* $\lambda \in]0,1[$ *the following estimates*

$$\text{there exists } \delta_i > 0 \text{ such that } \exp(x_i(t)) > \delta_i, \quad t\in[0,\omega], \quad i=1,\dots,5, \tag{45}$$

$$\text{there exists } \rho_i > 0 \text{ such that } \int_0^\omega \exp(x_i(t))dt < \rho_i, \quad i=1,\dots,5, \tag{46}$$

$$\text{there exists } d_i > 0 \text{ such that } \int_0^\omega \left|\frac{dx_i}{dt}(t)\right| dt < d_i, \quad i=1,\dots,5, \tag{47}$$

are satisfied. In particular, $\max_{t\in[0,\omega]}\exp(x_i(t)) \le \rho_i(\omega)^{-1}\exp(d_i)$ *and* $x_i(t) < \ln(\rho_i/\omega)+d_i$ *for* $t\in[0,\omega]$ *and* $i=1,\dots,5$.

Proof. We get the proof by application of Lemmas 4, 5 and 6, and the hypotheses in Equation (15). We notice that we can prove some relations in Equations (45)–(47) by a straightforward consequence of Lemmas 4, 5 and 6. More precisely, we can deduce

$$\text{(45) for } i=2,4,5, \text{ with } \delta_2=\exp\Big(-(\alpha+\chi+\mu)^\top\omega\Big),\ \delta_4=\exp\Big(-\mu^\top\omega\Big),\ \delta_5=\delta_4; \tag{48}$$

$$\text{(46) for } i=1,2,3,4, \text{ with } \rho_1=\frac{\omega\bar{b}}{\mu^\perp},\ \rho_2=\frac{\omega\bar{b}}{(\alpha+\chi+\mu)^\perp},\ \rho_3=\frac{\alpha^\top\rho_2}{(\gamma+\varepsilon+\mu)^\perp},\ \rho_4=\frac{\omega\bar{b}}{\phi^\perp}; \tag{49}$$

$$\text{(47) for } i=2,3,4, \text{ with } d_2=2\omega(\alpha+\chi+\mu)^\top,\ d_3=2\Big(\omega(\gamma+\varepsilon+\mu)^\top+\delta^\top\rho_4\Big),\ d_4=2\omega\bar{\mu}; \tag{50}$$

from Equations (28)–(30); (33)–(36); and (40)–(42); respectively. Meanwhile, to prove the remaining inequalities we proceed as follows:

(i) we prove that $\max_{t\in[0,\omega]}\exp(x_i(t)) \le \rho_i(\omega)^{-1}\exp(d_i)$ for $i=2,3,4$;
(ii) we prove Equation (45) for $i=1,3$;
(iii) we prove Equation (47) for $i=1$;

(*iv*) we prove Equation (46) for $i = 5$;
(*v*) we prove Equation (47) for $i = 5$.

Proof of (i). From Equation (49) and the intermediate value for integrals we can deduce that there exist $\xi_i \in [0, \omega]$ satisfying the inequality $x_i(\xi_i) < \ln(\rho_i/\omega)$ for $i = 2, 3, 4$. Then, by the fundamental theorem of calculus and Equation (50), we deduce that

$$x_i(t) = x_i(\xi_i) + \int_{\xi_i}^{t} \frac{dx_i}{dt}(t)dt < \ln(\rho_i/\omega) + \int_{\xi_i}^{t} \frac{dx_i}{dt}(t)dt < \ln(\rho_i/\omega) + d_i, \quad i = 2, 3, 4,$$

for any $t \in [0, \omega]$, which clearly implies *(i)*.

Proof of (ii). We notice that the assertion proved in *(i)* for $i = 2, 4$ and Equation (31) imply that

$$(pb)^{\perp} \leq \left[\mu^{\top} + \beta^{\top}\rho_2(\omega)^{-1}\exp(d_2)\right]\exp(x_1(t)) + \phi^{\top}\rho_4(\omega)^{-1}\exp(d_4), \tag{51}$$

for any $t \in [0, \omega]$. By hypotheses in Equation (15) we have that $(pb)^{\perp} - \phi^{\top}\rho_4(\omega)^{-1}\exp(d_4) \geq \kappa_1$, then Equation (51) implies Equation (45) for $i = 1$ with $\delta_1 = \kappa_1[\mu^{\top} + \beta^{\top}\rho_2(\omega)^{-1}\exp(d_2)]^{-1}$. Now, from the assertion proved in *(i)* for $i = 4$ and Equation (32) we can deduce Equation (45) for $i = 3$ with $\delta_3 = \alpha^{\perp}\left[(\gamma + \varepsilon + \mu)^{\top} + \delta^{\top}\rho_4(\omega)^{-1}\exp(d_4)\right]^{-1}$.

Proof of (iii). From Equation (48) and Lemma 6, we can follow that Equation (47) for $i = 1$ is satisfied with $d_1 = 2\omega\bar{b}/\delta_1$.

Proof of (iv). Form similar arguments and notation to the proof of step *(i)*, Equation (47) and Equation (47) for $i = 1$, we can deduce that

$$x_1(t) = x_1(\xi_1) + \int_{\xi_1}^{t} \frac{dx_1}{dt}(t)dt < \ln(\rho_1/\omega) + \int_{\xi_1}^{t} \frac{dx_1}{dt}(t)dt < \ln(\rho_1/\omega) + d_1,$$

for some $\xi_1 \in [0, \omega]$ and any $t \in [0, \omega]$. Then, $\max_{t\in[0,\omega]}\exp(x_1(t)) \leq \rho_1(\omega)^{-1}\exp(d_1)$. Now, from Equation (37) and the assertion proved in *(i)* for $i = 3$ we deduce that

$$\begin{aligned}\int_0^{\omega}\exp(x_5(t))dt &\leq \frac{\omega\bar{b}\phi^{\top}}{\mu^{\perp}\phi^{\perp}}\max_{t\in[0,\omega]}\exp(x_1(t)) + \frac{\varepsilon^{\top}\max_{t\in[0,\omega]}\exp(x_3(t))}{\mu^{\perp}\exp(K(0) - \mu^{\top}\omega)}\int_0^{\omega}\exp(x_5(t))dt \\ &\leq \frac{\omega\bar{b}\phi^{\top}\rho_1}{\mu^{\perp}\phi^{\perp}\omega}\exp(d_1) + \frac{\varepsilon^{\top}\rho_3\exp(d_3)}{\mu^{\perp}\omega\exp(-\mu^{\top}\omega)}\int_0^{\omega}\exp(x_5(t))dt. \end{aligned} \tag{52}$$

Thus, the hypotheses in Equation (15) implies

$$\kappa_2\int_0^{\omega}\exp(x_5(t))dt \leq \left(1 - \frac{\varepsilon^{\top}\rho_3\exp(d_3)}{\mu^{\perp}\omega\exp(-\mu^{\top}\omega)}\right)\int_0^{\omega}\exp(x_5(t))dt \tag{53}$$

$$\leq \frac{\omega\bar{b}\phi^{\top}\rho_1}{\mu^{\perp}\phi^{\perp}\omega}\exp(d_1), \tag{54}$$

which implies Equation (46) for $i = 5$ with $\rho_5 = \omega\bar{b}\phi^{\top}\rho_1\exp(d_1)\left[\mu^{\perp}\phi^{\perp}\omega\kappa_2\right]^{-1}$.

Proof of (iv). From *(i)* with $i = 4$ and Equation (48)

$$\int_0^{\omega} \left| \frac{dx_5}{dt}(t) \right| dt < 2\omega\overline{\mu} \max_{t\in[0,\omega]} \exp(x_4(t) - x_5(t))$$
$$\leq \frac{2\overline{\mu}\rho \exp(d_4)}{d_5} := d_5.$$

Then, Equation (47) for $i = 5$ is satisfied.

Summarizing we have that Equation (45) is followed by Equation (48) and *(ii)*; Equation (46) is a consequence of Equation (49) and *(iv)*; and Equation (47) is proved from Equation (50), and *(iii)* and *(v)*. Moreover, we observe that a sequence of similar arguments and notation to the proof of step *(i)*, Equations (47) and (47) for $i = 5$, implies that

$$x_5(t) = x_5(\xi_5) + \int_{\xi_5}^{t} \frac{dx_5}{dt}(t)dt < \ln(\rho_5/\omega) + \int_{\xi_5}^{t} \frac{dx_5}{dt}(t)dt < \ln(\rho_5/\omega) + d_5,$$

for some $\xi_5 \in [0, \omega]$ and any $t \in [0, \omega]$. Then, $\max_{t\in[0,\omega]} \exp(x_5(t)) \leq \rho_5(\omega)^{-1} \exp(d_5)$. Then, we get the additional and particular inequalities are followed from *(i)* and *(iv)*. □

3.2. Proof of (a)

We can prove the estimate in Equation (17) by application of Lemma (7).

3.3. Proof of (b)

If $\mathbf{x} \in \text{Ker } L$, then by the results of Section 2.3, we have that $\mathbf{x}(t) \in \mathbb{R}^5$ is constant for any $t \in [0, \omega]$. By notational convenience we consider that $\mathbf{x}(t) = (S_0, E_0, I_0, K_0, R_0)$. Then, from Equation (24) the condition $QN(\mathbf{x}^T) = QN((S_0, E_0, I_0, K_0, R_0)^T) = 0$ implies that

$$\begin{aligned}
0 &= \overline{pb} \exp(-S_0) - \overline{\beta} \exp(E_0) - \overline{\phi} \exp(K_0 - S_0) - \overline{\mu}, && \text{(55a)}\\
0 &= \overline{qb} \exp(-E_0) + \overline{\beta} \exp(S_0) - \overline{\alpha} - \overline{\chi} - \overline{\mu}, && \text{(55b)}\\
0 &= \overline{\alpha} \exp(E_0 - I_0) - \overline{\delta} \exp(K_0) - \overline{\gamma} - \overline{\varepsilon} - \overline{\mu}, && \text{(55c)}\\
0 &= \overline{\delta} \exp(I_0) + \overline{\gamma} \exp(I_0 - K_0) + \overline{\chi} \exp(E_0 - K_0) - \overline{\mu}, && \text{(55d)}\\
0 &= \overline{\phi} \exp(S_0 + K_0 - R_0) + \overline{\varepsilon} \exp(I_0 - R_0) - \overline{\mu}. && \text{(55e)}
\end{aligned}$$

Then, from Equation (55) and following similar arguments to the proof of Lemma 7, we can deduce that in this case an inequality of the type in Equation (46) is also valid, i.e.,

$$\exp(S_0) < \frac{\rho_1}{\omega}, \quad \exp(E_0) < \frac{\rho_2}{\omega} \quad \exp(I_0) < \frac{\rho_3}{\omega}, \quad \exp(K_0) < \frac{\rho_4}{\omega} \quad \text{and} \quad \exp(R_0) < \frac{\rho_5}{\omega}.$$

which implies Equation (18). Moreover, from Lemma 7 and the fact that Ker $L \subset$ Dom L, we can deduce that

$$\exp(S_0) > \delta_1, \quad \exp(E_0) > \delta_2, \quad \exp(I_0) > \delta_3, \quad \exp(K_0) > \delta_4, \quad \text{and} \quad \exp(R_0) > \delta_5.$$

Thus, the inequality in Equation (19) is also satisfied.

4. Proof of Theorem 3

4.1. A Previous Lemma

Lemma 8. *Let X and Y be the spaces defined on Equation (13); $\Omega \subset X$ the open ball centered at $(0,0,0,0,0)$ with radius*

$$h = \sum_{i=1}^{3} \max\left\{ \left| \ln(\delta_i) \right|, \left| \ln\left(\frac{\rho_i}{\omega_i}\right) \right| + d_i \right\}, \tag{56}$$

where δ_i, ρ_i and d_i are defined in the proof of Lemma 7; and L, N and Q the operators defined on Equations (5), (6) and (21), respectively. If Equation (15) is satisfied, the operators L and N satisfy the properties (C_1)–(C_3) of Theorem 5.

Proof. We prove (C_1) and (C_2) by contradiction argument and we prove (C_3) by application of invariance property of the topological degree. Indeed, we have that

(C_1) Let us assume that there are $\delta \in]0,1[$ and $\mathbf{x} \in \partial\Omega \cap \mathrm{Dom}\ L$ such that $L\mathbf{x} = \delta N\mathbf{x}$. Then, by application of Theorem 2-(a) we deduce that $\mathbf{x} \in \mathrm{Int}\ \Omega$ which is a contradiction to the assumption that $\mathbf{x} \in \partial\Omega$.

(C_2) Let us assume that there is $\mathbf{x} \in \partial\Omega \cap \mathrm{Ker}\ L$ such that $QN\mathbf{x} = 0$. Then, by application of Theorem 2-(b) we deduce that $\mathbf{x} \in \mathrm{Int}\ \Omega$ which is a contradiction to the assumption that $\mathbf{x} \in \partial\Omega$.

(C_3) Let us define the mapping $\Phi : \mathrm{Dom}\ L \times [0,1] \to X$ by the following relation

$$\Phi(\mathbf{x}, v) = \begin{bmatrix} \overline{pb}\exp(-x_1) - \overline{\beta}\exp(x_2) - \overline{\phi}\exp(x_4 - x_1) - \overline{\mu} \\ \overline{qb}\exp(-x_2) - \overline{\alpha + \chi + \mu} \\ \overline{\alpha}\exp(x_2 - x_3) - \overline{\delta}\exp(x_4) - \overline{\gamma + \varepsilon + \mu} \\ \overline{\delta}\exp(x_3) + \overline{\chi}\exp(x_2 - x_4) - \overline{\mu} \\ \exp(x_1 + x_4 - x_5) - \overline{\mu} \end{bmatrix} + v \begin{bmatrix} 0 \\ \overline{\beta}\exp(x_1) \\ 0 \\ \overline{\gamma}\exp(x_3 - x_4) \\ \overline{\varepsilon}\exp(x_3 - x_5) \end{bmatrix}.$$

We prove that $\Phi(\mathbf{x}, v) \neq 0$ when $\mathbf{x}^T \in \partial\Omega \cap \mathrm{Ker}\ L$ and $v \in [0,1]$. From Lemma 1 we recall that $\mathbf{x}^T(t) = (S_0, E_0, I_0, K_0, R_0) \in \mathbb{R}^5$ is a constant. Let us consider that the conclusion is false, then the constant vector $(S_0, E_0, I_0, K_0, R_0)^T$ with $\|(S_0, E_0, I_0, K_0, R_0)\| = h$ satisfies $\Phi(S_0, E_0, I_0, K_0, R_0, v) = 0$, that is,

$$\begin{aligned} 0 &= \overline{pb}\exp(-S_0) - \overline{\beta}\exp(E_0) - \overline{\phi}\exp(K_0 - S_0) - \overline{\mu}, \\ 0 &= \overline{qb}\exp(-E_0) - \overline{\alpha + \chi + \mu} + v\overline{\beta}\exp(S_0), \\ 0 &= \overline{\alpha}\exp(E_0 - I_0) - \overline{\delta}\exp(K_0) - \overline{\gamma + \varepsilon + \mu}, \\ 0 &= \overline{\delta}\exp(I_0) + \overline{\chi}\exp(E_0 - K_0) - \overline{\mu} + v\overline{\gamma}\exp(I_0 - K_0), \\ 0 &= \overline{\phi}\exp(S_0 + K_0 - R_0) - \overline{\mu} + v\overline{\varepsilon}\exp(I_0 - R_0). \end{aligned}$$

Then, by following similar reasoning steps to the proof of Theorem 2-(a) we get that $\|(S_0, E_0, I_0, K_0, R_0)^T\| < h$, which contradicts to the assumption that $\|(S_0, E_0, I_0, K_0, R_0)^T\| = h$. Let us consider $J = I : \mathrm{Im}\ Q \to \mathrm{Ker}\ L$ such that $\mathbf{x}^T \mapsto \mathbf{x}^T$, then by applying the Homotopy Invariance Theorem of Topology Degree, using the fact that the system

$$\begin{aligned} 0 &= \overline{pb}\exp(-x_1(t)) - \overline{\beta}\exp(x_2(t)) - \overline{\phi}\exp(x_4(t) - x_1(t)) - \overline{\mu}, \\ 0 &= \overline{qb}\exp(-x_2(t)) + \overline{\beta}\exp(x_1(t)) - \overline{\alpha + \chi + \mu}, \\ 0 &= \overline{\alpha}\exp(x_2(t) - x_3(t)) - \overline{\delta}\exp(x_4(t)) - \overline{\gamma + \varepsilon + \mu}, \\ 0 &= \overline{\delta}\exp(x_3(t)) + \overline{\chi}\exp(x_2(t) - x_4(t)) + \overline{\gamma}\exp(x_3(t) - x_4(t)) - \overline{\mu}, \\ 0 &= \overline{\phi}\exp(x_1(t) + x_4(t) - x_5(t)) + \overline{\varepsilon}\exp(x_3(t) - x_5(t)) - \overline{\mu}. \end{aligned}$$

has a unique solution $\mathbf{x}^{\star T} \in \partial\Omega \cap \text{Ker } L$, noticing that the determinant of the Jacobian of Φ at $\mathbf{x}^{\star T}$ is given by

$$\left|J_\Phi(\mathbf{x}^{\star T})\right| = -\left[\overline{\phi}\exp(x_1^\star + x_4^\star - x_5^\star) + \overline{\varepsilon}\exp(x_3^\star - x_5^\star)\right]\left(\Pi_1 + \Pi_2\right),$$

with Π_1 and Π_2 the positive functions

$$\begin{aligned}
\Pi_1 =& \left[-\overline{\beta}\exp(x_2^\star) - \mu\right]\left[-\overline{qb}\exp(x_2^\star)\right]\left[\overline{\alpha}\exp(x_2^\star - x_3^\star)\left(\overline{\chi}\exp(x_2^\star - x_4^\star) + \overline{\gamma}\exp(x_3^\star - x_4^\star)\right)\right. \\
&\left. + \overline{\delta}\exp(x_4^\star)\left(\overline{\delta}\exp(x_3^\star) + \overline{\gamma}\exp(x_3^\star - x_4^\star)\right)\right] \\
\Pi_2 =& -\overline{\beta}\exp(x_1^\star)\left\{\left[-\overline{\beta}\exp(x_2^\star)\right]\left[\overline{\alpha}\exp(x_2^\star - x_3^\star)\left(\overline{\chi}\exp(x_2^\star - x_4^\star) + \overline{\gamma}\exp(x_3^\star - x_4^\star)\right)\right.\right. \\
&\left. + \overline{\delta}\exp(x_4^\star)\left(\overline{\delta}\exp(x_3^\star) + \overline{\gamma}\exp(x_3^\star - x_4^\star)\right)\right] + \left[-\overline{\phi}\exp(x_2^\star - x_1^\star)\right] \\
&\left.\left[\overline{\alpha}\exp(x_2^\star - x_3^\star)\left(\overline{\delta}\exp(x_3^\star) + \overline{\gamma}\exp(x_3^\star - x_4^\star)\right) + \overline{\chi\alpha}\exp(2x_2^\star - x_3^\star - x_4^\star)\right]\right\},
\end{aligned}$$

and by Definition 3, we have that

$$\deg\left(JQN(\mathbf{x}^T, \Omega \cap \text{Ker } L, \mathbf{0}^T\right) = \deg\left(\Phi(\mathbf{x}, 1), \Omega \cap \text{Ker } L, \mathbf{0}^T\right) = \text{sgn}\left|J_\Phi(\mathbf{x}^{\star T})\right| = -1.$$

Hence, we get that $\deg(JQN, \Omega \cap \text{Ker } L, 0) \neq 0$ and prove that (C3) is valid.

Therefore, the assertions on items (C_1)-(C_3) of the Theorem 5 are valid for the given operators. □

4.2. *Proof of Theorem 3*

By Lemmata 7 and 8, we notice that the assumptions of the Theorem 5 are satisfied. Thus, there exist at least one solution of operator equation in Equation (12) belong $\text{Dom } L \cap \overline{\Omega} \subset X$, which implies the existence of at least one $\omega-$periodic solution of the system in Equation (4).

5. Proof of Theorem 4

The proof of Theorem 4 is a consequence of Theorems 3 and 1. Indeed, from Theorem 3 we deduce that there exists at least one $\omega-$periodic solution of Equation (4). Then, we get the proof of Theorem 4 by application of Theorem 1.

6. An Example

Let us consider that

$$\begin{aligned}
b(t) &= 100 + \cos(\pi t), & p(t) &= \frac{2}{3}\left(1 + \frac{1}{2}\sin(\pi t)\right), \\
q(t) &= 1 - p(t), & \alpha(t) &= 1.1960e - 90\sin^2\left(\frac{\pi}{2}t\right), \\
\beta(t) &= \cos^2\left(\frac{\pi}{2}t\right), & \phi(t) &= 1.1 + 1.0e - 10\sin(\pi t), \\
\mu(t) &= 1.0e - 15(1 + \sin(\pi t)), & \gamma(t) &= \frac{1}{2}(1.1 + 1.0e - 10\sin(\pi t)), \\
\chi(t) &= \frac{1}{4}(1.1 + 1.0e - 10\sin(\pi t)), & \delta(t) &= \frac{3}{4}(1.1 + 1.0e - 10\sin(\pi t)), \\
\epsilon(t) &= \frac{1.0e - 15}{4}(1 + \sin(\pi t)),
\end{aligned} \tag{57}$$

which are 2-periodic functions. We notice that

$$(pb)^{\perp} - \phi^{\top}(\phi^{\perp})^{-1}\overline{b}\exp(2\omega\overline{\mu}) \approx 0.15,$$
$$1 - \frac{\varepsilon^{\top}\alpha^{\top}\overline{b}}{\mu^{\top}(\gamma+\varepsilon+\mu)^{\perp}(\alpha+\chi+\mu)^{\perp}} \exp\left(\omega\left[(\gamma+\varepsilon+\mu)^{\top} + \phi^{\top}(\phi^{\perp})^{-1}\overline{b} + \mu^{\top}\right]\right) \approx 0.857,$$

and we have that the hypothesis in Equation (15) is satisfied by selecting $\kappa_1 \in]0, 0.15[$ and $\kappa_2 \in]0, 0.857[$. Thus, by application of Theorem 4, we deduce that the system in Equation (2) with coefficients defined by Equation (57) has at least one positive 2-periodic solution.

Author Contributions: Conceptualization, A.C. and F.H.; methodology, A.C.; investigation, all authors; writing—original draft preparation, F.N.-M. and E.L.; writing—review and editing, E.L.; funding acquisition, I.H. and F.N.-M. All authors have read and agreed to the published version of the manuscript.

Funding: A.C. and F.N.-M. thanks the support of research by projects DIUBB GI 172409/C, DIUBB 183309 4/R and DIUBB 192408 2/R at Universidad del Bío-Bío, Chile.

Acknowledgments: A.C. and E.L. thank the suggestions of colleges at Ciencias Básicas of Universidad del Bío-Bío. I.H. thanks to the grants program "Becas de doctorado" of ANID-Chile, 2017-Fol. 21171196.

Conflicts of Interest: The authors declare that they have no competing interests.

References

1. Anderson, R.M.; May, R.M. Population biology of infectious diseases I. *Nature* **1979**, *280*, 361–367 [CrossRef]
2. Hethcote, H. Qualitative analyses of communicable disease models. *Math. Biosci.* **1976**, *28*, 335–356 [CrossRef]
3. Hethcote, H.W. The mathematics of infectious diseases. *SIAM Rev.* **2000**, *42*, 599–653. [CrossRef]
4. Kuznetsov, Y.A.; Piccardi, C. Bifurcation analysis of periodic SEIR and SIR epidemic models. *J. Math. Biol.* **1994**, *32*, 109–121. [CrossRef] [PubMed]
5. Li, F.; Zhao, X.Q. A periodic SEIRS epidemic model with a time-dependent latent period. *J. Math. Biol.* **2019**, *78*, 1553–1579. [CrossRef]
6. Ma, W.; Song, M.; Takeuchi, Y. Global stability of an SIR epidemic model with time delay. *Appl. Math. Lett.* **2004**, *17*, 1141–1145 [CrossRef]
7. Martcheva, M. *An Introduction to Mathematical Epidemiology*; Texts in Applied Mathematics, 61; Springer: New York, NY, USA, 2015.
8. Smith, R. (Ed.) *Mathematical Modelling of Zombies*; University of Ottawa Press: Ottawa, ON, Canada, 2014.
9. Smith, R. *Modelling Disease Ecology with Mathematics*, 2nd ed.; AIMS Series on Differential Equations & Dynamical Systems, 5; American Institute of Mathematical Sciences (AIMS): Springfield, MO, USA, 2017.
10. Brauer, F.; Castillo-Chavez, C. *Mathematical Models in Population Biology and Epidemiology*, 2nd ed.; Texts in Applied Mathematics, 40; Springer: New York, NY, USA, 2012.
11. Kermack, W.O.; McKendrick, A.G. Contributions to the mathematical theory of epidemics: Part I. *Proc. R. Soc. A.* **1927**, *115*, 700–721.
12. Capasso, V. *Mathematical Structures of Epidemic Systems*; Lecture Notes in Biomathematics, 97; Springer-Verlag: Berlin, Germany, 2008.
13. Coronel, A.; Huancas, F.; Pinto, M. Sufficient conditions for the existence of positive periodic solutions of a generalized nonresident computer virus model. *Quaest. Math.* **2019**. [CrossRef]
14. Muroya, Y.; Enatsu, Y.; Li, H. Global stability of a delayed SIRS computer virus propagation model. *Int. J. Comput. Math.* **2014**, *91*, 347–367. [CrossRef]
15. Muroya, Y.; Kuniya, T. Global stability of nonresident computer virus models. *Math. Methods Appl. Sci.* **2015**, *38*, 281–295. [CrossRef]
16. Muroya, Y.; Li, H.; Kuniya, T. On global stability of a nonresident computer virus model. *Acta Math. Sci. Ser. B Engl. Ed.* **2014**, *34*, 1427–1445. [CrossRef]
17. Murray, W.H. The application of epidemiology to computer viruses. *Comput. Secur.* **1988**, *7*, 130–50. [CrossRef]
18. Ren, J.; Yang, X.; Zhu, Q.; Yang, L.-X.; Zhang, C. A novel computer virus model and its dynamics. *Nonlinear Anal. Real World Appl.* **2012**, *13*, 376–384. [CrossRef]

19. Alonso-Quesada, S.; De la Sen, M.; Ibeas, A. On the discretization and control of an SEIR epidemic model with a periodic impulsive vaccination. *Commun. Nonlinear Sci. Numer. Simul.* **2017**, *42*, 247–274. [CrossRef]
20. Ávila-Vales, E.; Rivero-Esquivel, E.; García-Almeida, G.E. Global dynamics of a periodic SEIRS model with general incidence rate. *Int. J. Differ. Equ.* **2017**. [CrossRef]
21. Bai, Z.; Zhou, Y. Global dynamics of an SEIRS epidemic model with periodic vaccination and seasonal contact rate. *Nonlinear Anal. Real World Appl.* **2012**, *13*, 1060–1068. [CrossRef]
22. Cooke, K.L.; van den Driessche, P. Analysis of an SEIRS epidemic model with two delays. *J. Math. Biol.* **1996**, *35*, 240–260. [CrossRef]
23. Du, Y.; Guo, Y.; Xiao, P. Freely-moving delay induces periodic oscillations in a structured SEIR model. *Internat. J. Bifur. Chaos Appl. Sci. Engrg.* **2017**, *27*, 1750122. [CrossRef]
24. Fang, H.; Wang, M.; Zhou, T. Existence of positive periodic solution of a hepatitis B virus infection model. *Math. Meth. Appl. Sci.* **2014**, *38*, 188–196. [CrossRef]
25. Gao, L.Q.; Mena-Lorca, J.; Hethcote, H.W. Four SEI endemic models with periodicity and separatrices. *Math. Biosci.* **1995**, *128*, 157–184. [CrossRef]
26. Huang, G.; Takeuchi, Y.; Maw, W.D. Global stability for delay SIR and SEIR epidemic models with nonlinear incidence rate. *Bull. Math. Biol.* **2010**, *72*, 1192–1207. [CrossRef] [PubMed]
27. Mateus, J.P.; Silva, C.M. Existence of periodic solutions of a periodic SEIRS model with general incidence. *Nonlinear Anal. Real World Appl.* **2017**, *34*, 379–402. [CrossRef]
28. Nakata, Y.; Kuniya, T. Global dynamics of a class of SEIRS epidemic models in a periodic environment. *J. Math. Anal. Appl.* **2010**, *363*, 230–237. [CrossRef]
29. Nistal, R.; de la Sen, M.; Alonso-Quesada, S.; Ibeas, A. Limit periodic solutions of a SEIR mathematical model for non-lethal infectious disease. *Appl. Math. Sci.* **2013**, *7*, 773–789. [CrossRef]
30. Yan, M.; Xiang, Z. Dynamic behaviors of an SEIR epidemic model in a periodic environment with impulse vaccination. *Discrete Dyn. Nat. Soc.* **2014**. [CrossRef]
31. Zhang, T.; Liu, J.; Teng, Z. Existence of positive periodic solutions of an SEIR model with periodic coefficients. *Appl. Math.* **2012**, *57*, 601–616. [CrossRef]
32. Zhang, Y.; Zhu, J. Dynamic behavior of an I2S2R rumor propagation model on weighted contract networks. *Phys. A* **2019**, *536*, 120981. [CrossRef]
33. Zhu, L.; Liu, M.; Li, Y. The dynamics analysis of a rumor propagation model in online social networks. *Phys. A* **2019**, *520*, 118–137. [CrossRef]
34. Gaines, R.; Mawhin, J. *Coincidence Degree and Nonlinear Diffrential Equations*; Springer: Berlin, Germany, 1977.

MDPI
St. Alban-Anlage 66
4052 Basel
Switzerland
Tel. +41 61 683 77 34
Fax +41 61 302 89 18
www.mdpi.com

Mathematics Editorial Office
E-mail: mathematics@mdpi.com
www.mdpi.com/journal/mathematics